Framework 8

MATHS C

David Capewell — Formerly Westfield School, Sheffield

Marguerite Comyns — Queen Mary's High School, Walsall

Gillian Flinton — All Saints Catholic High School, Sheffield

Paul Flinton — Chaucer School, Sheffield

Geoff Fowler — Maths Strategy Manager, Birmingham

Derek Huby — Mathematics Consultant, West Sussex

Peter Johnson — Wellfield High School, Leyland, Lancashire

Penny Jones — Waverley School, Birmingham

Jayne Kranat — Langley Park School for Girls, Bromley

Ian Molyneux — St. Bedes RC High School, Ormskirk

Peter Mullarkey — Netherhall School, Maryport, Cumbria

Nina Patel — Ifield Community College, West Sussex

OXFORD
UNIVERSITY PRESS

OXFORD
UNIVERSITY PRESS

Great Clarendon Street, Oxford OX2 6DP

Oxford University Press is a department of the University of Oxford.
It furthers the University's objective of excellence in research,
scholarship, and education by publishing worldwide in

Oxford New York

Auckland Cape Town Dar es Salaam Hong Kong Karachi
Kuala Lumpur Madrid Melbourne Mexico City Nairobi
New Delhi Shanghai Taipei Toronto

With offices in

Argentina Austria Brazil Chile Czech Republic France Greece
Guatemala Hungary Italy Japan Poland Portugal Singapore
South Korea Switzerland Thailand Turkey Ukraine Vietnam

Oxford is a registered trade mark of Oxford University Press
in the UK and in certain other countries

© Capewell et al. 2003

British Library Cataloguing in Publication Data

Data available

ISBN 978 0 19 914855 4

10 9 8 7

Typeset by Mathematical Composition Setters Ltd.

Printed in China by Printplus

Acknowledgements

The photograph on the cover is reproduced courtesy of Graeme Peacock

The publisher and authors would like to thank the following for permission
to use photographs and other copyright material: Science photo library,
page 1, Arcaid, page 15, Corbis UK, pages 72 and 92, Cordon Art, M. C. Escher,
'Sky and Water', page 119, Photodisc, page 128, Empics, page 220.

Figurative artwork by Paul Daviz.

About this book

This book has been written specifically for Year 8 of the Framework for Teaching Mathematics. It is aimed at students who are following the Year 8 teaching programme from the Framework.

The authors are experienced teachers and maths consultants who have been incorporating the Framework approaches into their teaching for many years and so are well qualified to help you successfully meet the Framework objectives.

The book is made up of units based on the sample medium term plans which complement the Framework document, thus maintaining the required pitch, pace and progression.

The units are:

Each unit comprises double page spreads that should take a lesson to teach. These are shown on the full contents list.

Problem solving is integrated throughout the material as suggested in the Framework.

How to use this book

This book is made up of units of work which are colour coded into: Number/Algebra (Purple), Algebra (Blue), Data (Pink), Number (Orange), Shape, space and measures (Green) and Problem solving (Light Green).

Each unit of work starts with an overview of the content of the unit, as specified in the Framework document, so that you know exactly what you are expected to learn.

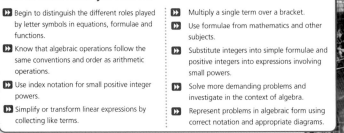

This unit will show you how to:

▶▶ Begin to distinguish the different roles played by letter symbols in equations, formulae and functions.

▶▶ Know that algebraic operations follow the same conventions and order as arithmetic operations.

▶▶ Use index notation for small positive integer powers.

▶▶ Simplify or transform linear expressions by collecting like terms.

▶▶ Multiply a single term over a bracket.

▶▶ Use formulae from mathematics and other subjects.

▶▶ Substitute integers into simple formulae and positive integers into expressions involving small powers.

▶▶ Solve more demanding problems and investigate in the context of algebra.

▶▶ Represent problems in algebraic form using correct notation and appropriate diagrams.

The first page of a unit also highlights the skills and facts you should already know and provides Check in questions to help you revise before you start so that you are ready to apply the knowledge later in the unit:

Inside each unit, the content develops in double page spreads which all follow the same structure.

The spreads start with a list of the learning outcomes and a summary of the keywords:

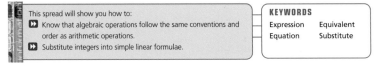

This spread will show you how to:
▶▶ Know that algebraic operations follow the same conventions and order as arithmetic operations.
▶▶ Substitute integers into simple linear formulae.

KEYWORDS

Expression   Equivalent
Equation   Substitute

The keywords are summarised and defined in a Glossary at the end of the book so you can always check what they mean.

Key information is highlighted in the text so you can see the facts you need to learn.

▶ Area of a rectangle = length × width

Examples showing the key skills and techniques you need to develop are shown in boxes. Also hint boxes show tips and reminders you may find useful:

example

A large bottle contains 2 litres of water.
How many pint glasses can you fill from the bottle?

....

1 litre ≃ 1.75 pints, so 2 litres ≃ 2 × 1.75 pints = 3.5 pints.
You can fill 3 pint glasses from the bottle.

Each exercise is carefully graded, set at three levels of difficulty:

▸ The first few questions provide lead-in questions, revising previous learning.
▸ The questions in the middle of the exercise provide the main focus of the material.
▸ The last few questions are challenging questions that provide a link to the Year 9 learning objectives.

At the end of each unit is a summary page so that you can revise the learning of the unit before moving on.

Check out questions are provided to help you check your understanding of the key concepts covered and your ability to apply the key techniques.

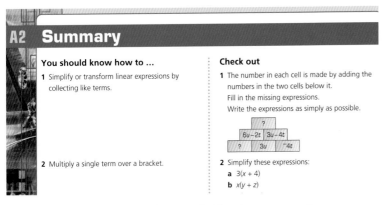

A2 Summary

You should know how to ...

1 Simplify or transform linear expressions by collecting like terms.

2 Multiply a single term over a bracket.

Check out

1 The number in each cell is made by adding the numbers in the two cells below it.
Fill in the missing expressions.
Write the expressions as simply as possible.

| ? |
| 6u − 2t | 3u − 4t |
| ? | 3u | ⁻4t |

2 Simplify these expressions:
a 3(x + 4)
b x(y + z)

The answers to the Check in and Check out questions are produced at the end of the book so that you can check your own progress and identify any areas that need work.

Contents

Numbers and sequences

This unit will show you how to:

- Add, subtract, multiply and divide integers.
- Use the sign change key on a calculator.
- Recognise and use multiples, factors, common factors, highest common factor, lowest common multiple and primes.
- Find the prime factor decomposition of a number.
- Use square, positive and negative square roots, cubes and cube roots, and index notation for small positive integer powers.
- Make and justify estimates and approximations.
- Generate and describe integer sequences.

- Generate terms of a linear sequence using term-to-term and position-to-term definitions.
- Begin to use linear expressions to describe the nth term of an arithmetic sequence.
- Solve more demanding problems and investigate in the context of number and algebra.
- Represent problems in algebraic form using correct notation.
- Choose efficient techniques for calculation.
- Suggest extensions to problems, conjecture and generalise.

You find patterns in nature – even in pine cones

Before you start

You should know how to ...

1 Order numbers.

2 Use simple tests of divisibility.
A number divides by:
- ▶ 2 if it is even
- ▶ 10 if it ends in 0
- ▶ 3 if the sum of the digits divides by 3
- ▶ 5 if it ends in a 5 or 0
- ▶ 4 if it divides by 2 twice

3 Describe sequences in words.
- ▶ You give the first term and the rule.

Check in

1 Put these numbers in order, highest first:
0.3   0.246   0.25   0.245

2 Here are four numbers:

| 30 | 60 | 15 | 20 |

Which of them divides by:
a 2   **b** 3
c 4   **d** 5

3 Describe these sequences in words:
a 2, 5, 8, 11, ...   **b** 3, 6, 12, 24, ...

This spread will show you how to:

▶▶ Add and subtract integers.

KEYWORDS

Integer	Positive
Digit	Compare
Negative	Sign change key

You use negative numbers to describe ...

... temperatures

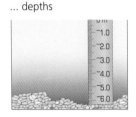

... the freezer is at a temperature of ⁻8.6 degrees.

... depths

... the river is ⁻6.28 m at its deepest point.

... measurements

... the diameter is within ±0.05 cm of 27 cm.

To compare negative numbers, first check the sign and then compare the place value of the digits.

Place these numbers in order, starting with the smallest:

⁻1.32, ⁻1.4, 0.6, ⁻1.35, ⁻1.3

⁻1.35 is further away from zero than ⁻1.32 so it is smaller.

Check the sign ...	⁻1.32, ⁻1.4, ⁻1.35, ⁻1.3, 0.6
Check the first decimal place ...	⁻1.4, ⁻1.32, ⁻1.3, ⁻1.35, 0.6
Check the second decimal place ...	⁻1.4, ⁻1.35, ⁻1.32, ⁻1.3, 0.6

You can check the order using a number line

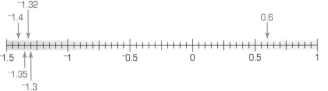

You can add and subtract negative integers.

▶ Adding a negative integer is the same as subtracting a positive integer.
▶ Subtracting a negative integer is the same as adding a positive integer.

Calculate   **a** $6 + 17 + {}^{-}4$ **b** $19 - {}^{-}4$

a $6 + 17 + {}^{-}4$
$= 6 + 17 - 4 = 19$

b $19 - {}^{-}4$
$= 19 + 4 = 23$

Exercise NA1.1

1 Order these temperatures from lowest to highest:

a $^-3\,°C$, $7\,°C$, $^-11\,°C$, $15\,°C$, $3\,°C$

b $^-12.6\,°C$, $8.3\,°C$, $0.2\,°C$, $^-4.28\,°C$, $4.3\,°C$, $^-8.35\,°C$, $^-8.4\,°C$, $13\,°C$, $4.38\,°C$

2 Find the number that lies exactly halfway between each of these pairs of temperatures:

a $2.8\,°C$ and $3.4\,°C$

b $^-3.5\,°C$ and $^-4.7\,°C$

c $^-2.6\,°C$ and $5.4\,°C$

d $^-4.8\,°C$ and $3.12\,°C$

e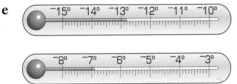

3

a A submarine is at a depth of 1854 m below sea level.
The bottom of the ocean is 2412 m below sea level.
How many metres must the submarine descend to reach the bottom of the ocean?

b Mr Skint is overdrawn by £48.32. He deposited £72.17 in his account. Can he now afford to buy a new pair of trainers at £23.99 without going overdrawn again?

4 Calculate these, deciding whether to use a mental or a written method:

a $7 + ^-4$ b $^-13 + 4 - 8$

c $-21 - ^-19$ d $265 + ^-307$

e $^-492 - 399$ f $^-427 + ^-318 - ^-453$

g $^-736 + 328$ h $^-834 + ^-256$

i $^-924 - ^-417$ j $473 - ^-261$

You can use the sign change key on your calculator to check your answers:

[7] [+] [4] [+/−] [=]

5 Investigation

In a Fibonacci number pattern each term is found by adding together the two previous terms, for example:

2	3	5	8	13	21

a Copy and complete these Fibonacci number patterns:

i

$^-8$	9				

ii

		9	6		

Write down what you notice about the last number in each pattern.

b Investigate all the pairs of integer starting numbers between $^-20$ and 20 that give a last number of 21.

?	?				21

Write down what you notice about the possible starting numbers.

6 Calculate these using an appropriate method.

a $^-4.3 + 2.7$ b $37 - ^-19 + 99$

c $42.1 + 36.7 + 12.4$

d $^-37.5 + ^-12.5$

e $143 + 16.8 - ^-4.9$

f $16.03 - 8.4 + 11.25 - ^-3.75$

g $^-\frac{3}{4} + \frac{2}{3}$ h $\frac{2}{5} - ^-\frac{3}{4}$

7 Puzzle

Here are six number cards:

You can add or subtract three of the numbers to make a target number:

$16.3 + -12.75 - 10.28 = -6.73$

Find the three numbers that make:

a a target number of $^-33.68$

b the largest target number

c the smallest target number.

Explain how you made your choice.

This spread will show you how to:
- ▶▶ Multiply and divide integers.
- ▶▶ Use the function key of a calculator for a sign change.

All the rules of multiplication and division apply to negative numbers but the sign may change in the answer.

▶ Multiplication is the same as repeated addition:

$$^-8 \times 4 = {^-8} + {^-8} + {^-8} + {^-8} = {^-32}$$
$$^-4 \times 8 = {^-4} + {^-4} + {^-4} + {^-4} + {^-4} + {^-4} + {^-4} + {^-4} = {^-32}$$

You can write times tables for negative integers ...

$^-8 \times 4 = {^-32}$	$^-4 \times 4 = {^-16}$
$^-8 \times 3 = {^-24}$	$^-4 \times 3 = {^-12}$
$^-8 \times 2 = {^-16}$	$^-4 \times 2 = {^-8}$
$^-8 \times 1 = {^-8}$	$^-4 \times 1 = {^-4}$

If you continue the pattern you can see what happens when you multiply a negative integer by a negative integer.

The answers become positive!

$^-8 \times 0 = 0$	$^-4 \times 0 = 0$
$^-8 \times {^-1} = 8$	$^-4 \times {^-1} = 4$
$^-8 \times {^-2} = 16$	$^-4 \times {-2} = 8$

You need to check the sign of the product when you multiply by a negative.

▶ Negative number × positive number = negative product
Negative number × negative number = positive product

$$13 \times 11 = 143$$
so
$$^-13 \times 11 = {^-143}$$
and
$$^-13 \times {^-11} = 143$$

Check your answers on your calculator using the ⊞ key.
To check $^-3 \times 2 = {^-6}$, press ③ ⊞ ✕ ② ＝

▶ Division is the inverse of multiplication:

$$^-8 \times 4 = {^-32} \quad \text{so} \quad ^-32 \div 4 = {^-8} \quad \text{and} \quad ^-32 \div {^-8} = 4$$
$$^-8 \times {^-4} = 32 \quad \text{so} \quad 32 \div {^-8} = {^-4} \quad \text{and} \quad 32 \div {^-4} = {^-8}$$

▶ Negative number ÷ positive number = negative quotient
Positive number ÷ negative number = negative quotient
Negative number ÷ negative number = positive quotient

Exercise NA1.2

1 Copy and complete these times tables:

a $5 \times 4 = 20$
  $5 \times 3 = 15$
  $5 \times 2 =$
  $5 \times 1 =$
  $5 \times 0 =$
  $5 \times =$
  $ \times =$

b $^-7 \times 2 =$
  $^-7 \times 1 =$
  $^-7 \times 0 =$
  $^-7 \times {}^-1 =$
  $^-7 \times =$
  $^-7 \times =$
  $ \times =$

2 Calculate these, deciding whether to use a mental or a written method:

a $^-8 \times {}^-9$
b $90 \div {}^-10$
c $12 \times {}^-9$
d $^-128 \div 4$
e 15×12
f $^-19 \times {}^-21$
g $208 \div {}^-16$
h $^-5 \times 28$
i $^-1.7 \times 100$
j $^-32 \div {}^-1000$

3 Puzzle
Copy and complete this multiplication grid:

×				
		27	⁻18	
4			24	
		⁻63		
⁻5	⁻40			

Write down any strategies you used to solve this problem.

4 Find the missing number in each of these calculations.
You should do all calculations mentally, but you may need to make some jottings.

a $12 \times \ ? \ = {}^-432$
b $? \times {}^-4 \ = \ {}^-42$
c $? \div 15 \ = \ {}^-6.2$
d $^-268 \div \ ? \ = \ 67$
e $24 \times \ ? \ = \ 720$
f $? \div {}^-3.2 \ = \ 70$

Check your answers using the sign change key on your calculator.

5 Puzzle
Find three consecutive integers with a product of ⁻336.

6 Investigation
Here is a grid of numbers:

⁻12	⁻8	⁻15
⁻3	5	14

A product pair can be made by multiplying a number by another number adjacent to it horizontally or vertically – not diagonally.
For example the product pairs for ⁻12 are:
$^-12 \times {}^-3 = 36$ or $^-12 \times {}^-8 = 96$

a Work out the seven different product pairs for this grid.
b What is the total of all your product pairs?
c Rearrange the numbers in the grid to make the largest total of product pairs.
d Explain why you think you have found the maximum product.

7 Challenge
Find the value of each of these expressions if $x = 5$, $y = {}^-4$ and $t = {}^-3$.

a **i** $2x$
  ii $4y$
  iii $2t + 6$
  iv $2x - 4y$
  v $x^2 - 2y$
  vi $\dfrac{3x}{t}$
  vii $\dfrac{3x}{y}$

b Make up five expressions of your own using x, y and t with a value of ⁻12.
For example: $4t = {}^-12$.

This spread will show you how to:

▶▶ Recognise and use multiples, factors, common factors, highest common factor, lowest common multiple and primes.

▶▶ Find the prime factor decomposition of a number.

KEYWORDS

Multiple   Factor   Prime

Highest common factor

Lowest common multiple

Prime factor decomposition

▶ A factor is a number that divides exactly into another number.

The factors of 28 are 1, 2, 4, 7, 14 and 28.

▶ A **prime factor** is a prime number that divides exactly into another number.

The prime factors of 28 are 2 and 7.

▶ Any number can be expressed as the product of its prime factors.

$28 = 2 \times 2 \times 7 = 2^2 \times 7$.

There are two common methods of expressing a number as a product of its prime factors.

Factor trees

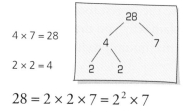

$4 \times 7 = 28$

$2 \times 2 = 4$

$28 = 2 \times 2 \times 7 = 2^2 \times 7$

Division by prime numbers

For example: to express 126 as the product of its prime factors:

Divide by 2: $126 \div ②= 63$

Divide by 3: $63 \div ③= 21$ (63 doesn't divide by 2)

Divide by 3: $21 \div ③= ⑦$ which is prime.

$126 = 2 \times 3 \times 3 \times 7$

$$
\begin{array}{r}
2 \enclose{)}{126} \\
3 \enclose{)}{\ 63} \\
3 \enclose{)}{\ 21} \\
7
\end{array}
$$

$126 = 2 \times 3 \times 3 \times 7$
$= 2 \times 3^2 \times 7$

You can use prime factors to find the **highest common factor (HCF)** and **lowest common multiple (LCM)** of 28 and 126:

List the prime factors: $28 = 2 \times 2 \times 7$ $126 = 2 \times 3 \times 3 \times 7$

$HCF = 2 \times 7 = 14$ $LCM = 2 \times 2 \times 7 \times 3 \times 3 = 252$

Use the common factors.   Use all the multiples of the first number and the extras from the second!

Exercise NA1.3

1 **Game** for 2 players
You need a 0–9 dice and a copy of this grid.

Take it in turns to roll the dice and write the number in one of the boxes of your grid. Once the boxes are full, you should each have five 2-digit numbers.
Work out all of the factors of each of your 2-digit numbers.
The player with the most factors is the winner.

2 Use factors where appropriate to calculate mentally:
 a 25 × 12
 b 18 × 15
 c 390 ÷ 15
 d 216 ÷ 72
 e 6.3 × 20
 f 14 ÷ 20

3 **a** Identify which of these numbers are prime numbers:

 117   143   173   323   571

 b Explain clearly the method you have used to identify a prime number.

4 Express these numbers as products of their prime factors:
 a 18   **b** 12
 c 24   **d** 45
 e 63   **f** 100
 g 180   **h** 576
 i 480   **j** 1080

5 **Investigation**
The number 20 can be written as
2 × 2 × 5.
It has three prime factors.
 a Find three numbers with exactly four prime factors.
 b Find the two 2-digit numbers with the largest number of prime factors.
 c Find the largest 3-digit number with exactly five prime factors.

6 **Investigation**
The numbers 12 and 30 have a HCF of 6 and a LCM of 60.
Investigate the connection between the HCF and LCM for different pairs of numbers.
Explain how the prime factors of a pair of numbers can help you find their HCF and LCM.

7 Find the HCF and LCM of:
 a 9 and 24   **b** 21 and 18
 c 40 and 56   **d** 48, 54 and 72
 e 112, 154 and 63   **f** 126 and 588
 g 425 and 816   **h** 1650 and 3465

8 **Puzzle**
 a Find all the pairs of numbers that have a HCF of 6 and an LCM of 252.
 b Use the digits 7, 3, 5, 6, 9 and 6 to make two 3-digit numbers with a HCF of 36.
 Explain the method you used to solve each puzzle.

9 **Challenge**
 a Find the greatest number that when divided into 423 and 885 leaves a remainder of exactly 3 in each case.
 b What is the smallest number that is divisible by 12, 18, 20 and 24?

This spread will show you how to:

▶▶ Use square, positive and negative square roots, cubes and cube roots, and index notation for small positive integer powers.

▶▶ Make and justify estimates and approximations of calculations.

KEYWORDS

Square number   Index
Cube number   Power
Square root   Cube root

You can write square numbers and cube numbers using index notation.

$$1^2 = 1 \times 1 = 1$$
$$2^2 = 2 \times 2 = 4$$
$$3^2 = 3 \times 3 = 9$$
$$4^2 = 4 \times 4 = 16$$
$$5^2 = 5 \times 5 = 25$$

$$1^3 = 1 \times 1 \times 1 = 1$$
$$2^3 = 2 \times 2 \times 2 = 8$$
$$3^3 = 3 \times 3 \times 3 = 27$$
$$4^3 = 4 \times 4 \times 4 = 64$$
$$5^3 = 5 \times 5 \times 5 = 125$$

▶ A positive integer has two square roots, one positive and one negative:

$\sqrt{16} = 4$ or $^-4$ because $4 \times 4 = 16$ and $^-4 \times {}^-4 = 16$

You usually use the positive root.

▶ You can find the square root of a square number using factors:

$\sqrt{1600} = \sqrt{(16 \times 100)} = \sqrt{16} \times \sqrt{100} = 4 \times 10 = 40$

You can use trial and improvement to estimate the square root of a number.

example

Find $\sqrt{20}$ to 1 decimal place.

You know that

$$4^2 = 16 \quad \text{and} \quad 5^2 = 25$$
$$\text{so } 4 < \sqrt{20} \text{ and } 5 > \sqrt{20}$$

Try 4.5: $4.5^2 = 20.25$ so $4.5 > \sqrt{20}$

Try 4.4: $4.4^2 = 19.36$ so $4.4 < \sqrt{20}$

4.5^2 is closer than 4.4^2 to 20
so, 4.5 is closer than 4.4 to $\sqrt{20}$

Check on a calculator:
$\sqrt{20} = 4.4721359 \ldots = 4.5$ (1 dp)

$\sqrt{20} \approx 4.5$ (1 decimal place)

▶ You can represent powers of any number using index notation:

$10^6 = 10 \times 10 \times 10 \times 10 \times 10 \times 10 = $ 'ten to the power of six'

$(^-5)^4 = {}^-5 \times {}^-5 \times {}^-5 \times {}^-5 = $ 'negative five to the power of four'

▶ A positive integer has a positive cube root.

For example, $^3\sqrt{1000} = 10$ because $10 \times 10 \times 10 = 1000$.

Exercise NA1.4

1 Calculate:

a 3^2 **b** 5^4 **c** 10^8 **d** $(^-2)^5$

e 0.1^3 **f** $\sqrt{81}$ **g** $\sqrt[3]{8}$ **h** 19^2

i 23^2 **j** $13^2 + 14^2$

2 Puzzle

Some numbers can be represented as the sum of two square numbers.

For example: $41 = 4^2 + 5^2$

Find all the numbers less than 100 that can be represented as the sum of two square numbers.

3 Use prime factor decomposition to:

a Write each of these numbers using index notation, for example: $25 = 5^2$

   i 243   **ii** 625   **iii** 256

   iv 289   **v** 529   **vi** 729

   vii 2197   **viii** 4913

b Find:

   i $\sqrt{2500}$ **ii** $\sqrt{196}$ **iii** $\sqrt{441}$

   iv $\sqrt{784}$ **v** $\sqrt{1225}$

4 Game for 2 players.

Shuffle two sets of 1–9 digit cards. Deal out five cards each. Place three of the remaining cards face up to represent the target number:

 represents a target number of 572.

Take it in turns to use two cards to make an expression involving a power that is as close as possible to the target number.

 represents the expression $5^4 = 625$

The player who is closest scores the target number.

Play the game four times.

5 Use trial and improvement to find each of these numbers to 1 decimal place.

a $\sqrt{40}$ **b** $\sqrt{65}$ **c** $\sqrt{130}$ **d** $\sqrt{200}$

Use the square root key on your calculator to check your answers.

6 Puzzle

Find two consecutive numbers with a product of 2862.

Explain your method for solving this problem.

7 Use your calculator to work out these problems:

a $\sqrt{10} = 3.162278$

Calculate $(3.162278)^2$

Explain why the answer is not 10.

b Estimate which of these numbers is greater:

   i 12^3 or 25^2 **ii** 0.12^3 or 0.25^2

Write down the methods you have used to estimate the answers.

Use your calculator to check your answers.

8 Investigation

a Work out:

$4^2 - 3^2$

$8^2 - 7^2$

$13^2 - 12^2$

Investigate the squares of other pairs of consecutive integers.

Explain anything you have noticed.

b Investigate the difference between other pairs of square numbers, for example:

$5^2 - 3^2$

$12^2 - 7^2$

Explain anything you have noticed.

This spread will show you how to:

▶▶ Generate and describe integer sequences.

▶▶ Generate terms of a sequence using a term-to-term definition.

KEYWORDS

Generate Flow chart

Sequence T(*n*)

Term

▶ A sequence is a set of numbers that follow a rule.

3, 6, 9, 12, 15, ... is the sequence of the multiples of three.

The first term is 3, the fourth term is 12.
You write T(1) = 3 and T(4) = 12.

You can find the next term from the previous term by adding 3.

The term-to-term
rule is 'add 3'.

You can generate a sequence of numbers using a flow chart.
In a flow chart, different-shaped boxes have different meanings:

Start or end Command Yes or No question

---

example

Generate a sequence from this flow chart:

Start

Write down
1

Add on
6

Write down
your answer

Is the answer
more than 25?

No

Yes

Stop

The flow chart gives this information:

The first term is 1: T(1) = 1

The term-to-term rule is 'add 6'.

So T(2) = T(1) + 6 = 1 + 6 = 7
    T(3) = T(2) + 6 = 7 + 6 = 13
    T(4) = T(3) + 6 = 13 + 6 = 19
    T(5) = T(4) + 6 = 19 + 6 = 25
    T(6) = T(5) + 6 = 25 + 6 = 31

The answer is now more than 25 so that is the end of the sequence.

The sequence is 1, 7, 13, 19, 25, 31.

Exercise NA1.5

1 Here are four flow charts. Follow the instructions to generate a sequence for each flow chart.

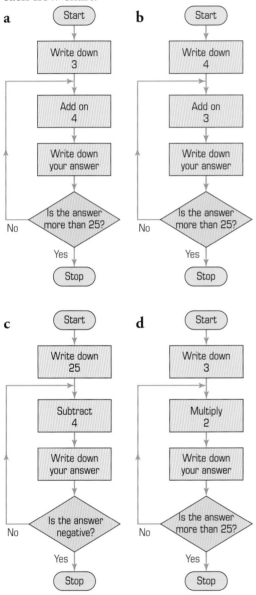

a

Start → Write down 3 → Add on 4 → Write down your answer → Is the answer more than 25? No / Yes → Stop

b

Start → Write down 4 → Add on 3 → Write down your answer → Is the answer more than 25? No / Yes → Stop

c

Start → Write down 25 → Subtract 4 → Write down your answer → Is the answer negative? No / Yes → Stop

d

Start → Write down 3 → Multiply 2 → Write down your answer → Is the answer more than 25? No / Yes → Stop

2 The terms of a sequence are generated by substituting values of n into the expression $(5n - 2)$.
Write down the first six terms of this sequence.
Substitute $n = 1$ for the first term, $n = 2$ for the second term and so on.

3 Draw a flow chart with instructions to generate the first six terms of the sequence in question **2**.

4 For each of these sequences, write down the first six terms.
   a The first term is 1, each term is double the previous term.
   b The first term is 24, each term is half the previous term.
   c Start at 0 and count forward by 1, 2, 3, 4 ...
   d Start at 1 and count forward by 3, 5, 7, 9, 11. How else could this sequence be generated?
   e The third term is 60 and each term is double the previous term.

5 **Investigation** – Growing rectangles
   a Draw the next pattern in this sequence.

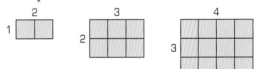

   b Find the area for each rectangle. Write down the first five terms of the sequence of areas.
   c Explain this sequence in words.
   d The 10th term in the sequence has 110 squares. Explain why.
   e Which term in the sequence has 72 squares?
   f Investigate the area of each rectangle in this sequence.

This spread will show you how to:

▶▶ Generate terms of a linear sequence using term-to-term and position-to-term definitions of the sequence.

▶▶ Begin to use linear expressions to describe the *n*th term of an arithmetic sequence.

KEYWORDS

*n*th term
Term-to-term rule
Position-to-term rule
General term
Linear sequence
Arithmetic sequence

▶ A linear sequence has a constant difference pattern:

This is a linear sequence as the difference is always 3:

$$8, \quad 11, \quad 14, \quad 17, \quad 20, ...$$

A linear sequence can also be called an arithmetic sequence

▶ You can find the general, *n*th term of a linear sequence using the common difference.

If the difference is 3, the *n*th term is $T(n) = 3n +$ something.

Compare the sequence with $3n$.

Sequence	8	11	14	17	20
$3n$	3	6	9	12	15
Difference	5	5	5	5	5

The term-to-term rule is +3.
The position-to-term rule is $3n + 5$.

The general term is $T(n) = 3n + 5$.

---

example

Here are the first four patterns in a sequence:

Find $T(n)$.

The number of lines in the four patterns are: 6   10   14   18
The difference is 4. Compare with $4n$:   4   8   12   16
   2   2   2   2

$T(n) = 4n + 2$.

---

example

In a linear sequence, $T(1) = 6$ and $T(3) = 12$.
Write down the first five terms and the *n*th term of this sequence.

$T(3) - T(1) =$ twice the difference $= 6$. So the difference is 3.
The first five terms are   6   9   12   15   18
Compare with $3n$:   3   6   9   12   15
   3   3   3   3   3

The *n*th term is $T(n) = 3n + 3$.

Exercise NA1.6

1 Write down the first six terms of each of these linear sequences.
   a $T(n)$ is $3n + 5$
   b $T(n)$ is $42 - 8n$
   c $T(1)$ and $T(2)$ are 3 and 9
   d $T(1)$ and $T(2)$ are 5 and 11
   e $T(3)$ is 9 and $T(4)$ is 12.

2 Here are the first three patterns of a sequence:

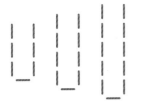

   a Draw the next two patterns.
   b List the numbers of lines in the first six patterns in the sequence.
   c Predict the number of lines in the 7th and 12th patterns in this sequence.
   d Explain how you know that the 100th term has 205 lines.
   e Write down the expression for the general term $T(n)$ for this sequence.

3 Here are the first three patterns of a sequence:

   a Draw the next pattern.
   b Write down the number of lines in the first six terms of the sequence.
   c Predict the number of lines in the 12th pattern of the sequence. Explain in words how you worked out your answer.
   d Explain how you know that the 100th pattern has 304 lines.
   e Write down the expression for the general term $T(n)$ for this sequence.

4 Write down the first six terms and the general term $T(n)$ of these linear sequences.
   a $T(1)$ and $T(2)$ are 12 and 16
   b $T(1)$ and $T(3)$ are 12 and 16
   c $T(3)$ and $T(5)$ are 12 and 16
   d $T(1)$ and $T(6)$ are 21 and $^-9$
   e $T(2)$ and $T(4)$ are 8 and 20
   f $T(3)$ and $T(6)$ are 18 and $^-3$

5 Jordan has put paving stones around her six rose beds. Here are three of them:

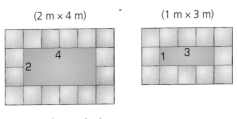

(2 m × 4 m)   (1 m × 3 m)

(1 m × 6 m)

   a Work out the number of 1 metre square slabs used for each border.
   b The other three rose beds measure 1 m × 2 m, 2 m × 3 m and 3 m × 4 m. How many square slabs are needed for each bed?
   c Jordan has created another flower bed which is a rectangle l metres × w metres. How many square slabs will she need to put a border around this?

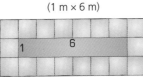

   d Check that your rule will work for the different-sized rose beds in parts **a** and **b**.

You should know how to ...

1 Add, subtract, multiply and divide integers.

2 Find the prime factor decomposition of a number.

3 Begin to use linear expressions to describe the *n*th term of an arithmetic sequence.

4 Represent problems in algebraic form using correct notation.

Check out

1 Calculate:
   a $^-243 + ^-175$
   b $^-243 - ^-175$
   c $^-12 \times 14$
   d $168 \div ^-14$

2 Write these numbers as a product of their prime factors:
   a 420
   b 630

3 Here is a pattern of lines:

   1st pattern

   2nd pattern

   3rd pattern

   4th pattern

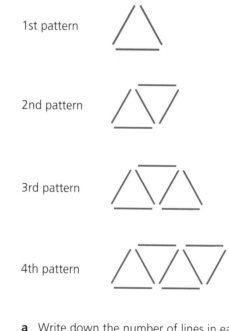

   a Write down the number of lines in each pattern.
   b Describe in words how the pattern grows.
   c Predict the number of lines in the 5th, 7th, and 10th patterns.
     Explain how you worked out your answer.

4 Write down an expression for the general term for the pattern of lines in question **3**.

This unit will show you how to:

▶▶ Identify alternate and corresponding angles.

▶▶ Understand a proof that:

▶ the sum of angles of a triangle is 180° and of a quadrilateral is 360°

▶ the exterior angle of a triangle is equal to the sum of the two interior opposite angles.

▶▶ Solve geometric problems using side and angle properties of equilateral, isosceles and right-angled triangles and special quadrilaterals, explaining reasoning with diagrams and text.

▶▶ Classify quadrilaterals by their geometric properties.

▶▶ Use straight edge and compasses to construct:

▶ the midpoint and perpendicular bisector of a line segment

▶ the bisector of an angle

▶ the perpendicular from a point to a line

▶ the perpendicular from a point on a line.

▶▶ Use logical argument to establish the truth of a statement.

▶▶ Identify exceptional cases or counter-examples.

Architects use angles and shapes to great effect.

Before you start

You should know how to ...

1 Classify angles.

▶ An acute angle is between 0° and 90°.

▶ An obtuse angle is between 90° and 180°.

▶ A reflex angle is between 180° and 360°.

2 Find missing angles on straight lines and corners.

▶ There are 90° on a corner.

▶ There are 180° on a straight line.

3 Solve simple equations.

Check in

1 **a** Draw two acute angles.

b Draw two obtuse angles.

c Draw two reflex angles.

2 Find the labelled angles.

3 Solve these equations:

a $3a = 180°$ **b** $2b = 90°$

c $a + 125° = 180°$ **d** $2a + 40° = 180°$

This spread will show you how to:
▶▶ Identify alternate and corresponding angles.
▶▶ Understand a proof that the sum of angles in a triangle is 180°.

KEYWORDS
Vertically opposite angles
Corresponding angles
Alternate angles
Proof   Prove

▶ When a line crosses parallel lines, eight angles are formed:

Remember:
Parallel lines never meet

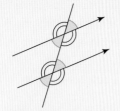

There are four acute angles and four obtuse angles.
The obtuse angles are all the same size.
The acute angles are all the same size.

Acute + obtuse = 180° as they are on a straight line.

Vertically opposite angles are equal.

Corresponding angles are equal.

Alternate angles are equal.

You can use these facts to prove that the sum of angles in a triangle is 180°.

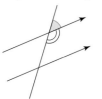

$$a + b + c = 180°$$

Proof
Add a line parallel to AB:

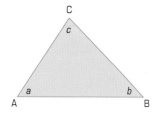

$x = a$ (alternate angles)
$y = b$ (alternate angles)
$x + c + y = 180°$ (angles on a straight line)

So $a + c + b = 180°$

Exercise S1.1

1 Find all the labelled angles, explaining your reasons.

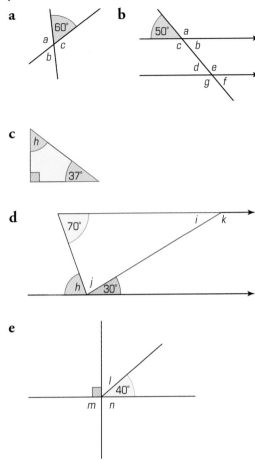

a

b

c

d

e

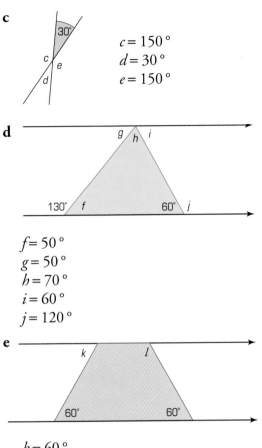

c

$c = 150°$
$d = 30°$
$e = 150°$

d

$f = 50°$
$g = 50°$
$h = 70°$
$i = 60°$
$j = 120°$

e

$k = 60°$
$l = 120°$

2 In these questions, explain why the answers given are correct.

The first one is done for you

a

$a = 140°$

$(a + 40° = 180°$, angles on a straight line add up to 180°)

b

$b = 50°$

3 Challenge

Prove that △ and ○ are equal.

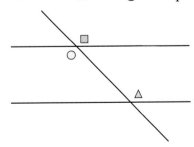

4 Challenge

Prove that the opposite angles of a parallelogram are equal.

17

This spread will show you how to:

⏩ Understand a proof that:

▶ the sum of angles in a quadrilateral is 360°.

▶ the exterior angle of a triangle is equal to the sum of the two interior opposite angles.

KEYWORDS
Quadrilateral   Pentagon
Interior angle   Polygon
Exterior angle
Supplementary angles
Complementary angles

▶ Any quadrilateral can be split into two triangles by drawing in a diagonal:

You can use this fact to show that the sum of interior angles in a quadrilateral is 360°:

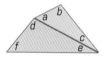

$a + b + c = 180°$ (angles in a triangle)
$d + e + f = 180°$ (angles in a triangle)

So $a + b + c + d + e + f = 360°$

Angles that add to 180° are called **supplementary angles**. Angles that add to 90° are called **complementary angles**.

▶ Any interior angle has an associated exterior angle:

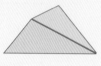

In this triangle:

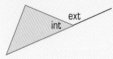

Interior + exterior = 180° (angles on a straight line)

Interior + $b + c$ = 180° (angles in a triangle)

So the exterior angle = $b + c$
                       = the sum of the other two interior angles

Exercise S1.2

1 Find ∠ABC.

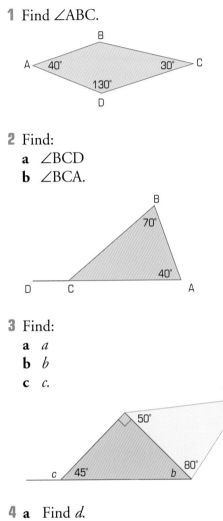

2 Find:
  a ∠BCD
  b ∠BCA.

3 Find:
  a *a*
  b *b*
  c *c*.

4 a Find *d*.

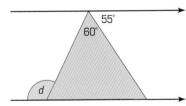

  b Find *e*.

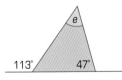

5 Find ∠DCB.

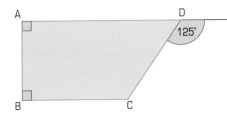

6 Find *f*.

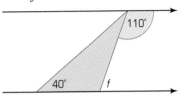

7 a A quadrilateral has angles of 110°, 90°, 70° and *x*.
  Find *x* and sketch the quadrilateral.
  b A quadrilateral has angles of *x*, 2*x*, 3*x* and 4*x*.
  Find *x* and sketch the shape.

8 Sketch the triangle ABC with ∠ABC = 120° and ∠BCA = 20°.

9 Use this diagram:

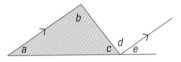

  to prove that *a* + *b* = *d* + *e*.

10 Prove that the angle sum of a pentagon is 540°.

11 Challenge
  Find a formula for the angle sum for an *n*-sided polygon.

19

This spread will show you how to:

⏩ Solve geometrical problems using side and angle properties of equilateral, isosceles and right-angled triangles, explaining reasoning with diagrams and text.

KEYWORDS
Equilateral   Scalene
Isosceles

▶ Triangles with special properties have special names:

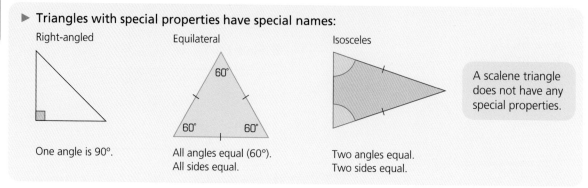

Right-angled

One angle is 90°.

Equilateral

60°

60°   60°

All angles equal (60°).
All sides equal.

Isosceles

Two angles equal.
Two sides equal.

A scalene triangle does not have any special properties.

You can use their special properties to help solve problems.

example

Find the missing angles in this triangle:

A   B
x   x   y

50°

C

Sum of angles in a triangle = 180°

$$x + x + 50° = 180°$$
$$2x = 130°$$
$$x = 65°$$

$$x + y = 180° \text{ (angles on a straight line)}$$
$$65° + y = 180°$$
$$y = 115°$$

example

Calculate the angles marked by letters, giving reasons for your answers.

46° a   x   x   b   36°
         120°

The shaded triangle is isosceles so the angles marked x are equal:

$$x + x + 120° = 180° \text{ (angles in a triangle)}$$
$$2x = 60°$$
$$x = 30°$$

$$46° + a + x = 180° \text{ (angles on a straight line)}$$
$$46° + a + 30° = 180°$$
$$a = 104°$$

$$x + 36° + b = 180° \text{ (angles on a straight line)}$$
$$30° + 36° + b = 180°$$
$$b = 114°$$

Exercise S1.3

1 Calculate the angles marked with letters.

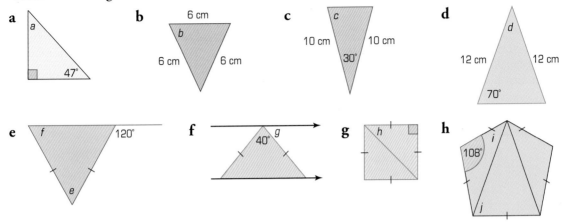

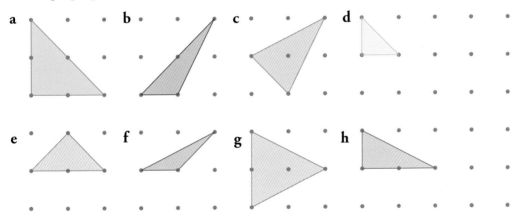

2 Describe each of these triangles, giving any special names, side and angle properties.

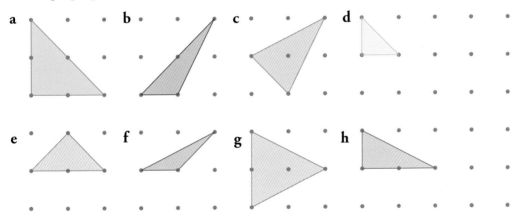

3 The diagram shows a regular hexagon.
The angle at each vertex is 120°.

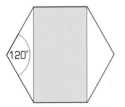

Explain why the shaded shape is a rectangle.

This spread will show you how to:

▶▶ Solve geometrical problems using side and angle properties of quadrilaterals, explaining reasoning with diagrams and text.

▶▶ Classify quadrilaterals by their geometric properties.

KEYWORDS

Quadrilateral   Parallelogram
Trapezium   Rhombus
Isosceles   Diagonal
Kite

▶ Special quadrilaterals have special properties:

Parallelogram

Opposite sides equal.
Opposite angles equal.

Rhombus

Four equal sides.
Diagonals bisect at 90°.

Kite

Adjacent sides equal.
Diagonals bisect at 90°.

Trapezium

One pair of parallel sides.

Isosceles trapezium

One pair of parallel sides.
Two pairs of equal angles.

Arrowhead (delta)

Two pairs of equal sides.
One pair of equal angles.

You can use these properties to solve problems.

example

Find the missing angles in these diagrams, giving reasons for your answers.

a 127°, a

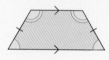
b d, b, 47°, c

c 62°, e

a $90° + 90° + 127° + a = 360°$ (angles in a quadrilateral)
$$307° + a = 360°$$
$$a = 53°$$

b
$$b = 47°$$ (opposite angles equal)
$$c = d$$ (opposite angles equal)
$$c + c + 47° + 47° = 360°$$ (angles in a quadrilateral)
$$c = d = 133°$$

c Diagonals bisect at 90°, so use this triangle:

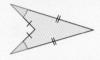
31°, e

$$e + 31° + 90° = 180°$$ (angles in a triangle)
$$e = 59°$$

Exercise S1.4

Calculate the angles marked with letters.
Give reasons for your answers.

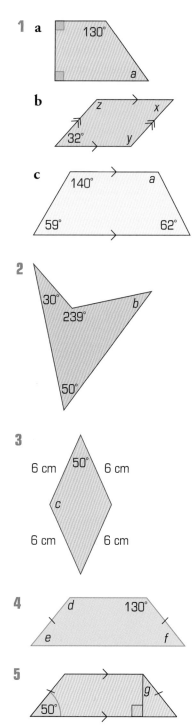

1 a 130° a

b z x 32° y

c 140° a 59° 62°

2 30° 239° b 50°

3 6 cm 50° 6 cm c 6 cm 6 cm

4 d 130° e f

5 50° g

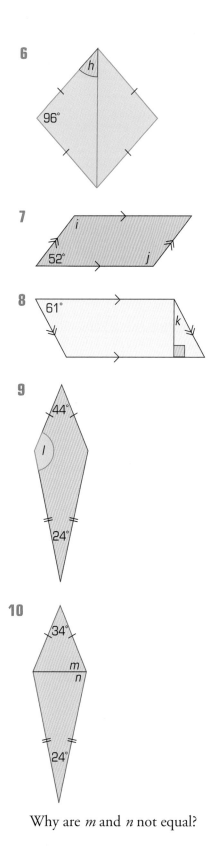

6 h 96°

7 i 52° j

8 61° k

9 44° l 24°

10 34° m n 24°

Why are *m* and *n* not equal?

Bisecting angles and lines

This spread will show you how to:

▶▶ Use straight edge and compasses to construct:
  ▶ The midpoint and perpendicular bisector of a line segment.
  ▶ The bisector of an angle.

KEYWORDS

Midpoint	Acute
Bisect	Obtuse
Bisector	Straight edge
Perpendicular	Compasses

▶ If you bisect a line or angle you cut it exactly in half:

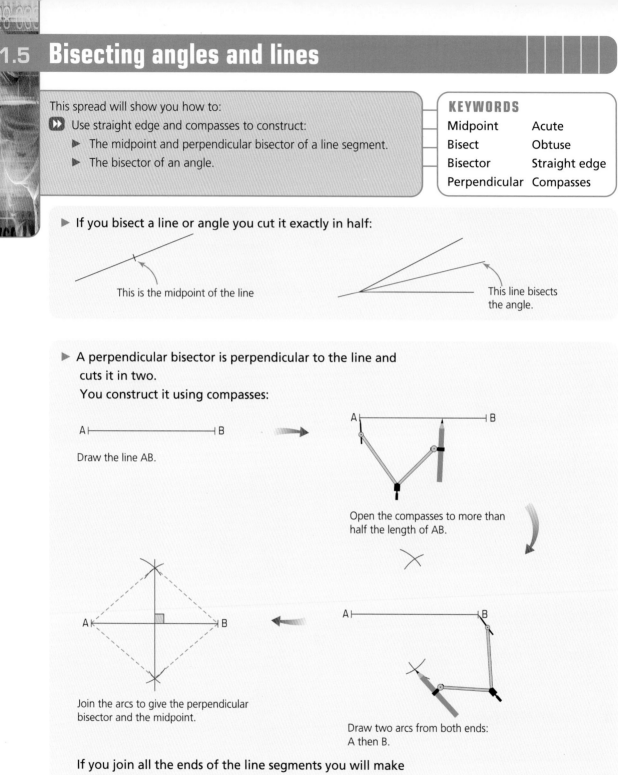

This is the midpoint of the line

This line bisects the angle.

▶ A perpendicular bisector is perpendicular to the line and cuts it in two.
  You construct it using compasses:

A ⊢————————————⊣ B

Draw the line AB.

A ⊢————————————⊣ B

Open the compasses to more than half the length of AB.

A ⊢————————————⊣ B

Draw two arcs from both ends: A then B.

A ⊢————————————⊣ B

Join the arcs to give the perpendicular bisector and the midpoint.

If you join all the ends of the line segments you will make a rhombus.

You can also use compasses to bisect an angle.
Question **4** will show you how.

Exercise S1.5

For all constructions, use a pencil and do **not** rub out your construction lines.

1   Draw a line segment between 8 cm and 12 cm and construct the perpendicular bisector.

2   Draw a line segment between 12 cm and 16 cm and construct the perpendicular bisector.

3   Construct a triangle.

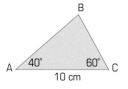

Construct the perpendicular bisector for each of the three sides.
What do you notice?

4   Draw an acute angle and label it XAY as shown.

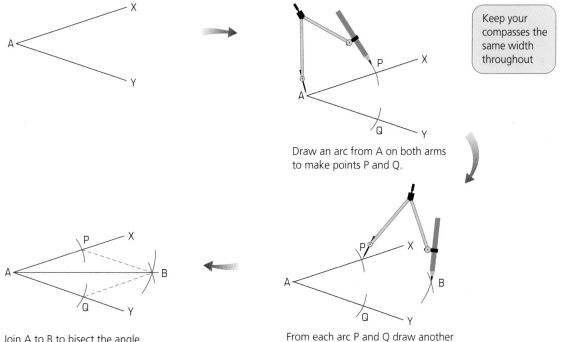

Keep your compasses the same width throughout

Draw an arc from A on both arms to make points P and Q.

From each arc P and Q draw another arc at B.

Join A to B to bisect the angle.

What is the name of the shape APBQ?

5   Repeat question **4** starting with an obtuse angle.

6   Draw any triangle.
Construct the angle bisector for each of the three angles.
What do you notice?

This spread will show you how to:

▶▶ Use straight edge and compasses to construct:
  ▶ The perpendicular from a point to a line.
  ▶ The perpendicular from a point on a line.

KEYWORDS
Perpendicular
Straight edge
Compasses

▶ The shortest distance from a point to a line is the perpendicular distance.

▶ You construct a perpendicular from a point to a line like this:

Open the compasses so that the distance is longer than the distance from the point to the line.

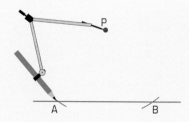

Construct two arcs A and B from the point to the line.

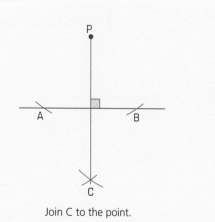

Join C to the point.

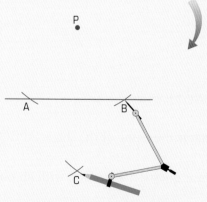

Keep the compasses the same width and construct an arc from A and from B to meet at C.

Exercise S1.6

1 Copy or trace the diagrams and construct the perpendicular from the point to the line.

a

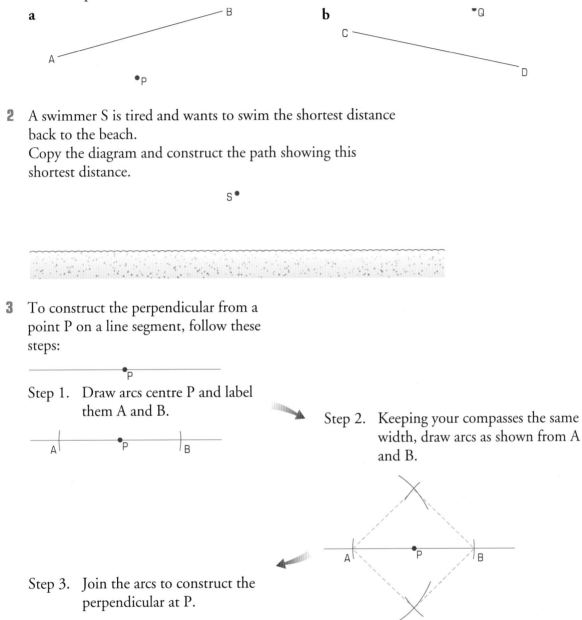

b

2 A swimmer S is tired and wants to swim the shortest distance back to the beach.
Copy the diagram and construct the path showing this shortest distance.

S•

3 To construct the perpendicular from a point P on a line segment, follow these steps:

Step 1.   Draw arcs centre P and label them A and B.

Step 2.   Keeping your compasses the same width, draw arcs as shown from A and B.

Step 3.   Join the arcs to construct the perpendicular at P.

4 Copy the diagrams. Construct the perpendicular from the point on the line segment.

a

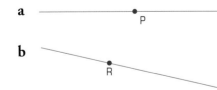

P

b

R

Summary

You should know how to …

1 Identify alternate and corresponding angles.

Check out

1 The diagram shows angles between parallel lines:

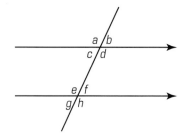

Write down:

a two pairs of alternate angles

b two pairs of vertically opposite angles

c two pairs of corresponding angles.

2 Understand a proof that the sum of angles in a triangle is 180° and in a quadrilateral is 360°.

2 a Use this diagram to explain why the sum of the angles in a triangle is 180°:

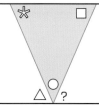

b Explain why the angle sum of this quadrilateral is 360°:

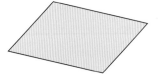

3 Use straight edge and compasses to construct:
- ▶ the midpoint and perpendicular bisector of a line segment
- ▶ the bisector of an angle
- ▶ the perpendicular from a point to a line
- ▶ the perpendicular from a point on a line.

3 a Draw a line 10 cm long.
Use a straight edge and compasses to construct the midpoint and perpendicular bisector of the line.

b Draw an acute angle and construct the bisector of your angle.

4 Use logical argument to establish the truth of a statement.

4 Prove that a triangle cannot have a reflex angle.

This unit will show you how to:

- Use the vocabulary of probability when interpreting the results of an experiment.
- Appreciate that random processes are unpredictable.
- Know that if the probability of an event occurring is p then the probability of it not occurring is $1 - p$.
- Find and record all possible mutually exclusive outcomes for single events in a systematic way, using diagrams and tables.

- Estimate probabilities from experimental data.
- Understand that:
    - if an experiment is repeated there may be, and usually will be, different outcomes
    - increasing the number of times an experiment is repeated generally leads to better estimates of probability.
- Solve problems and investigate in the context of probability.
- Identify the necessary information to solve a problem.

There's a 15% chance of rain today so you'd better take your umbrella.

Probability is a measure of chance.

Before you start

You should know how to ...

1 Use the vocabulary of probability.

2 Use the probability scale from 0 to 1.

```
0                              1
├──────────────────────────────┤
Impossible                Certain
```

3 Find and justify simple probabilities based on equally likely outcomes.

Check in

1 Write down an event which is:
  a impossible   **b** unlikely
  c highly likely   **d** definite.

2 Mark these probabilities on a scale.
  0.1   0.5   0.9
  Describe each one in words.

3 When a fair dice is rolled, what is the probability that the score is:
  a 5   **b** even   **c** prime?

This spread will help you to:

▶▶ Use the vocabulary of probability when interpreting the results of an experiment.

▶▶ Appreciate that random processes are unpredictable.

▶▶ Know that if the probability of an event occurring is p, then the probability of it not occurring is $1 - p$.

When you carry out a trial, there will be a particular outcome.

The set of all the possible outcomes is called the sample space.

The outcome is what actually occurs.

In a trial of tossing a coin, there are two outcomes in the sample space: Heads and Tails.

In a trial of rolling a dice, there are 6 outcomes in the sample space: **1**, **2**, **3**, **4**, **5** and **6**.

An event is a collection of possible outcomes.

When the outcomes are equally likely, the probability of an event is:

▶ $\text{Probability} = \dfrac{\text{Number of favourable outcomes}}{\text{Total number of outcomes in sample space}}$

The result will be between 1 (certain) and 0 (impossible).

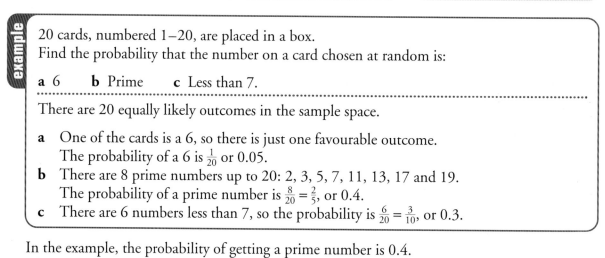

example

20 cards, numbered 1–20, are placed in a box.

Find the probability that the number on a card chosen at random is:

a 6   **b** Prime   **c** Less than 7.

There are 20 equally likely outcomes in the sample space.

a One of the cards is a 6, so there is just one favourable outcome.
The probability of a 6 is $\frac{1}{20}$ or 0.05.

b There are 8 prime numbers up to 20: 2, 3, 5, 7, 11, 13, 17 and 19.
The probability of a prime number is $\frac{8}{20} = \frac{2}{5}$, or 0.4.

c There are 6 numbers less than 7, so the probability is $\frac{6}{20} = \frac{3}{10}$, or 0.3.

In the example, the probability of getting a prime number is 0.4.
The probability of *not* getting a prime number is $1 - 0.4 = 0.6$.

▶ If the probability of an event occurring is p, the probability of it *not* occurring is $1 - p$.

Exercise D1.1

1 For each trial, list the possible outcomes in the sample space.

   a Peter puts three balls, coloured red, green and blue, in a bag.
      He chooses a ball at random, and records its colour.

   b Mary picks a playing card at random from an ordinary pack of 52 cards. She makes a note of the suit of the chosen card.

   c Paul keeps a weather diary as part of a project. Every morning at 8am he looks out of the window, and records whether or not it is raining.

   d Peter, Mary and Paul want to decide who will do a job. They each write their name on a piece of paper. They put the pieces of paper in a hat and pick one at random.

   e Phones 4 Free picks a person at random from their computer database.
      If the person chosen is aged 12 or older, they win a mobile phone.

2 For each part of question **1**, explain whether or not all of the outcomes in the sample space are equally likely.

3 For each trial, list the possible outcomes in the sample space.
Put a ring around the favourable outcomes.

   a Shane rolls an ordinary dice in a board game.
      He needs a score of 4 to finish.

   b Kelvin picks a digit card at random, from a pack numbered 0 to 9.
      He wins a prize if the number chosen is a multiple of 3.

   c A game uses a pack of 12 cards, one for each month of the year.
      Players choose a card at random, and win a prize if the name of the month chosen contains the letter B.

4 **a** Use your answer to question **3a** to copy and complete this calculation:

> Probability that Shane finishes the game
> $$= \frac{\text{Number of favourable outcomes}}{\text{Total number of outcomes}}$$

   b Do a similar calculation for the other parts of question **3**.

5 The table shows the probabilities for some events. Copy the table, and complete it to show the probability of each event *not* occurring. The first one is done for you.

Event	Probability that the event occurs	Probability that the event does *not* occur
A	$\frac{3}{5}$	$\frac{2}{5}$
B	0.28	
C	67%	
D	$\frac{4}{9}$	
E	0.525	

6 A trial has three possible outcomes.
The table shows some of the probabilities:

Outcome	Probability of the outcome occurring	Probability of the outcome *not* occurring
A	0.35	
B		
C	0.28	

Copy and complete the table.

7 To win a prize at a fairground stall you must score five or more with one dart.
Probability of scoring 0 or 1 $= \frac{2}{20}$
Probability of scoring 2 or 3 $= \frac{6}{20}$
Probability of scoring 4 $= \frac{7}{20}$
What is the probability of winning a prize?

This spread will show you how to:
▶▶ Find and record all possible mutually exclusive outcomes for single events and two successive events in a systematic way, using diagrams and tables.

KEYWORDS
Event   Outcome
Sample space diagram

You can list all the possible outcomes of a trial in simple cases.
For example, the possible outcomes when rolling a dice are: 1, 2, 3, 4, 5 and 6.

A sample space diagram can help you list all the outcomes correctly.

example

A blue dice and a red dice are rolled, and the scores are added. What are the possible outcomes?

The sample space diagram shows all the possible outcomes.

There are 36 possible outcomes, which give 11 different scores.

For example, the event 'the total score is 7' can occur in 6 different ways.

+	1	2	3	4	5	6
1	2	3	4	5	6	7
2	3	4	5	6	7	8
3	4	5	6	7	8	9
4	5	6	7	8	9	10
5	6	7	8	9	10	11
6	7	8	9	10	11	12

Red dice score (columns), Blue dice score (rows)

A sample space diagram is a systematic way of listing the outcomes.

example

Josh chooses two flavours of ice cream to make up a double cone. There are three flavours that he can choose from. How many selections could he make?

Call the three flavours A, B and C. The entry 'BA' in the table means that the first scoop is flavour B and the second scoop is flavour A.
The sample space diagram shows the nine different selections that Josh could make.

	A	B	C
A	AA	AB	AC
B	BA	BB	BC
C	CA	CB	CC

Scoop 2 (columns), Scoop 1 (rows)

In the example 'AB' is the same ice cream as 'BA', but you count it twice – you choose a different flavour first.

Exercise D1.2

1 The scores on two five-sided spinners are *multiplied* together.

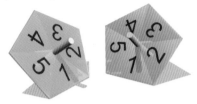

Copy and complete the sample space diagram to show all the possible outcomes.

		Score on red spinner				
	x	1	2	3	4	5
Score on blue spinner	1	1		3		
	2					
	3					
	4		8		16	
	5					

2 The Monte Carlo Café sells sandwiches with a choice of three types of bread:

> brown, white or rye

and three different fillings:

> egg, tuna or cheese.

Copy and complete the sample space diagram to show all the different types of sandwich that are available.

		Filling		
		Egg	Tuna	Cheese
Bread	Brown			
	White			
	Rye		TR	

> TR means tuna on rye

3 The Copenhagen Café has four choices of bread and five choices of filling for its sandwiches.

a *Without* drawing a sample space diagram, write down how many different types of sandwich are available at the Copenhagen Café.

b Explain how you would work out the number of different types of sandwich for *any* number of types of bread and filling.

4 A trifle is made with two layers of jelly. There are three flavours to choose from: raspberry, strawberry and tangerine. Copy and complete the diagram to show the different outcomes that are possible.

		Top layer		
		Raspberry	Strawberry	Tangerine
Bottom layer	Raspberry	RR		
	Strawberry			TS
	Tangerine			

5 Draw tables to show all the possible outcomes for the following situations:

a Light bulbs are available in strengths of 40 W, 60 W or 100 W. They can be clear or frosted.

b A company makes house signs in large, medium and small sizes. They are available in black, silver or gold.

c A mother gives birth to twins. (Each twin is either a boy or a girl.)

6 A pizza company sells pizzas with six different toppings. You must choose two different toppings for a pizza.

a How many possible outcomes are there when there are six toppings to choose from?

b Investigate for other numbers of toppings.

More theoretical probability

This spread will show you how to:

▶▶ Find and record all possible mutually exclusive outcomes for single events and two successive events in a systematic way.

KEYWORDS

Sample space diagram
Equally likely
Favourable outcome

You can use sample space diagrams to help calculate probabilities.

▶ You use the formula:

$$\text{Probability} = \frac{\text{Number of favourable outcomes}}{\text{Total number of outcomes}}$$

example

A coin is flipped twice.
What is the probability that the two results will be the same?

The sample space diagram shows the possible outcomes.
There are 4 possible outcomes, as shown in the table.
There are 2 favourable outcomes (HH and TT).

Using the formula:

Probability that both results are the same $= \frac{2}{4} = \frac{1}{2} = 0.5$

		Second toss	
		Head	Tail
First toss	Head	HH	HT
	Tail	TH	TT

Remember that to use the formula, all the outcomes must be equally likely.

example

In a game at a school fair, you spin a fair four-sided spinner and roll an ordinary dice.
The scores are multiplied together.
You win if the result is an odd number.
What is the probability of winning?

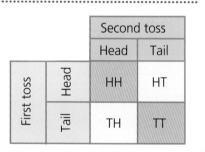

The sample space diagram shows that there are 24 outcomes.
There are 6 odd results.

Probability of winning $= \frac{6}{24} = \frac{1}{4} = 0.25$

	×	Dice score				
	×	1	2	3	4	5	6
Spinner score	1	1	2	3	4	5	6
	2	2	4	6	8	10	12
	3	3	6	9	12	15	18
	4	4	8	12	16	20	24

Exercise D1.3

1 The sample space diagram shows the possible outcomes when two ordinary dice are rolled and their scores are added.

		Score on first dice				
		1	2	3	4	5	6
Score on second dice	1	2	3	4	5	6	7
	2	3	4	5	6	7	8
	3	4	5	6	7	8	9
	4	5	6	7	8	9	10
	5	6	7	8	9	10	11
	6	7	8	9	10	11	12

Use the diagram to find the probability that:
a The total of the scores is 4
b The total is less than 9
c The scores on the two dice are equal
d The total is 1
e Both scores are odd

2 Two packs of 52 playing cards are shuffled and placed face down on a table. The top cards are turned over.
The table shows the possible outcomes.

		Pack 1			
		♣	♦	♥	♠
Pack 2	♣	♣♣	♦♣	♥♣	♠♣
	♦	♣♦	♦♦	♥♦	♠♦
	♥	♣♥	♦♥	♥♥	♠♥
	♠	♣♠	♦♠	♥♠	♠♠

Find the probability that:
a Both cards are the same suit
b One card is a club, and the other is a heart
c Both cards are the same colour
d Neither card is a diamond
e Exactly one of the cards is a heart

3 Use appropriate sample space diagrams to answer each of these problems.
a A fair five-sided spinner and an ordinary dice are dropped onto a table.

What is the probability that the dice shows a higher score than the spinner?
b Two ordinary dice are rolled, and the scores are multiplied.
Find the probability that the product will be less than 10.
c There are three male puppies and two females in a litter. A vet chooses a puppy at random for a dental check, and then chooses a puppy for a blood test.
What is the probability that the puppies chosen are of the same sex?

4 A fair five-sided spinner is marked with the letters A to E.
A player spins the spinner twice.
If the same letter comes up both times, the player wins.
a What is the probability of winning?
b Investigate the probability of winning with spinners with different numbers of sides.

5 Challenge
Karen rolls a red dice and a blue dice. She divides the score on the red dice by the score on the blue dice.
Find the probability that the answer is a whole number.

This spread will help you to
▶▶ Estimate probabilities from experimental data.
▶▶ Understand that increasing the number of times an experiment is repeated generally leads to better estimates of probability.

KEYWORDS
Secondary data   Estimate
Trial
Experimental probability

You need to estimate probabilities using experimental data:

▶ When the outcomes are not equally likely
▶ When the sample space is too large to list.

The outcomes of a game are rarely equally likely.

example

Estimate the probability that a Premiership football match will result in:
a A home win   **b** An away win   **c** A draw

You can estimate the probabilities from secondary data.

Results of 100 Premiership football matches

Result	Tally	Frequency
Home win	卌 卌 卌 卌 卌 卌 卌 卌 卌 l	46
Away win	卌 卌 卌 卌 卌 卌	30
Draw	卌 卌 卌 卌 llll	24

There is no reason to believe that the three outcomes in this case are equally likely.

Based on these figures, you can estimate the probabilities as:
a 46% for a home win
b 30% for an away win
c 24% for a draw.

▶ To estimate the probability of an event, use the formula:

$$\text{Probability} = \frac{\text{Number of successes}}{\text{Total number of trials}}$$

'Number of successes' means the number of times that the event occurs.

The larger the number of trials, the more reliable the estimate.

example

Find the probability that a word selected from *Pride and Prejudice* contains the letter 'e'.

Using the first 10 words,   probability $= \frac{3}{10} = 0.3$

Using the first 100 words,   probability $= \frac{39}{100} = 0.39$

Using the first chapter,   probability $= \frac{335}{853} = 0.393$ (3 dp)

1 In each of these situations explain whether you would find the probabilities by:

▶ calculating a theoretical probability based on equally likely outcomes, or
▶ carrying out an experiment to estimate the probabilities.

a 100 raffle tickets are placed in a box, and a winning ticket is drawn at random.
b There will be snow in London on Christmas Day.
c The next car could be a red one.
d Mrs Joy's new baby will be a girl.
e Robbie picks a book at random from a shelf. Some of the books are fiction, some are non-fiction.
f A plant bought from a garden centre may or may not survive in a garden.

2 Sparko matchboxes are supposed to contain 50 matches each. Here are the number of matches contained in a sample of 10 boxes.

52	48	53	50	49
55	54	50	51	53

Use the data in the table to estimate the probability that a box of Sparko matches will contain fewer than 50 matches.

3 The heights in cm of 20 visitors to a theme park are shown in the table.

142	110	187	159	190
162	155	173	148	126
113	148	109	134	159
163	165	179	177	144

Visitors must be at least 140 cm tall to go on the Banshee ride. Estimate the probability that a visitor picked at random will be tall enough for the ride.

4 A police patrol records the speeds of vehicles in three different locations.

Location	Speed limit (mph)	Recorded speeds (mph)				
A	30	28	32	41	29	25
B	50	55	49	63	61	55
C	70	92	68	73	75	81

a For each location, estimate the probability that a car selected at random is breaking the speed limit.
b What could the police patrol do if they wanted more reliable estimates of these probabilities?

5 The table shows the number of goals scored by St Peter's Football Club in each match last season.

2	4	2	3	4	5
3	6	8	0	6	2
3	2	5	5	3	4
0	3	6	2	5	5
2	9	6	2	5	6
1	1	5	1	4	2

Use the data in the table to estimate the probability that in their next match they will score:

a 4 goals or more
b No goals
c 6 goals
d 7 goals

Write a paragraph about your answers. Explain how reliable you think the probabilities you calculated are.

Probability and variation

This spread will help you to:
- ▶▶ Appreciate that random processes are unpredictable.
- ▶▶ Understand that if an experiment is repeated there may be, and generally will be, different outcomes.

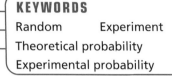

KEYWORDS

Random   Experiment
Theoretical probability
Experimental probability

Probabilities describe how likely things are to happen in the long term. Actual outcomes can vary a lot in the short term.

example

Explain why these statements are false:

a

> There must be something wrong with this dice – I've rolled it 12 times, and I still haven't got a 6!

b

> I've tossed this coin 3 times, and got 3 heads. It's bound to be tails next time.

a   Although you would *expect* to get a 6, it is possible to go 12 rolls without one. If you went 20 rolls without success, you should start to get suspicious.

b   The probability that you get tails next time is still 0.5 – the coin has not got a memory!

An experiment is a series of trials. Each time you carry out an experiment, you usually expect to get different results.

> The probability stays the same, but the results vary.

example

Place seven red marbles and three green marbles in a bag.

- ▶ Choose a marble at random, note its colour, then put it back.
- ▶ Repeat this four more times.

Estimate the probability of choosing a green marble.

The table shows the results of a computer simulation of this experiment.

Trial	1	2	3	4	5	6	7	8	9	10
Green	1	0	1	2	0	2	1	3	1	1
Red	4	5	4	3	5	3	4	2	4	4

4 out of 5 of the marbles were red on the first trial.

None of the numbers were green in the fifth trial.

The theoretical probability of choosing a green is $\frac{3}{10} = 0.3$
The experimental probability of choosing a green is $\frac{12}{50} = 0.24$
which is close to the theoretical value even though the results varied.

Exercise D1.5

1 Experiment

Roll an ordinary dice.

- ▶ If the result is a six, stop.
- ▶ If not, roll again.
- ▶ Keep rolling until you get a six.
- ▶ Write down the number of rolls you needed each time.

Carry out the experiment at least 10 times – 20 times is better.

2 Dubious data

Here are five sets of data from the Experiment in question **1**.
In each set, the experiment was repeated 20 times.

A	10	5	1	2	3	4	10	3	8	3	1	5	11	16	3	1	1	1	8	1
B	2	5	1	3	1	1	1	6	11	5	2	2	5	2	5	2	2	8	1	2
C	5	4	7	8	7	6	5	5	5	6	8	6	8	7	6	5	7	5	6	5
D	4	8	10	11	8	11	43	29	10	6	35	2	43	10	11	1	1	7	22	11
E	3	8	1	5	5	3	2	6	5	5	7	24	12	2	9	18	1	4	5	6

Three of the sets are genuine experimental data.
The other two sets are fakes.
Explain which sets of data you think are fake.
Give a reason for your answer.

3 Game

This is a dice game for two players.
The first player to bank 10 points is the winner.

Use the data from the experiment in question 1 to help you play the game.

- ▶ Player 1 rolls a dice.
- ▶ If the dice shows a six, player 1 scores nothing and it is player 2's turn.
- ▶ If the dice shows anything other than a six, player 1 scores 1 point.
- ▶ Player 1 can roll the dice as many times as they like, and will score 1 point each time, as long as they don't get a 6 on the dice. At any time, player 1 can say 'bank' instead of rolling the dice.
  When this happens:
  - ▶ The score for this go is added to Player 1's banked total.
  - ▶ It is then player 2's go to roll the dice.

Remember – only banked scores count towards the winning total!

Comparing experiment and theory

This spread will help you:
- ▶▶ Estimate probabilities from experimental data.
- ▶▶ Understand that increasing the number of times an experiment is repeated generally leads to better estimates of probability.

KEYWORDS
Experiment   Simulation
Theoretical probability
Experimental probability
Biased   Theory

Computers can be used to simulate a large number of trials.

The scores on two dice are added. The sample space diagram shows the 36 possible outcomes.

The score 8 appears 5 times, so the theoretical probability of getting a score of 8 is $\frac{5}{36}$.

The table shows the theoretical probability of each score.

		White dice score					
		1	2	3	4	5	6
Red dice score	1	2	3	4	5	6	7
	2	3	4	5	6	7	8
	3	4	5	6	7	8	9
	4	5	6	7	8	9	10
	5	6	7	8	9	10	11
	6	7	8	9	10	11	12

Score	2	3	4	5	6	7	8	9	10	11	12
Probability	$\frac{1}{36}$	$\frac{2}{36}$	$\frac{3}{36}$	$\frac{4}{36}$	$\frac{5}{36}$	$\frac{6}{36}$	$\frac{5}{36}$	$\frac{4}{36}$	$\frac{3}{36}$	$\frac{2}{36}$	$\frac{1}{36}$

If you roll the dice 36 times, you would expect to get these results:

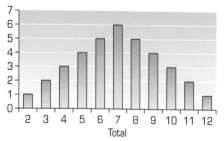

A computer simulation gives these results:

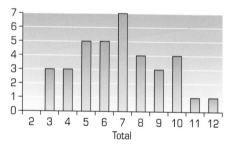

If you increase the number of trials to 360, instead of 36, the results change:

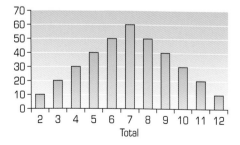

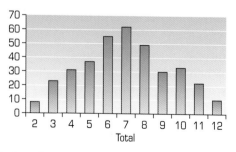

The theoretical distribution is the same shape, but the frequencies are 10 times as big.

This time, the experimental distribution is a much closer fit to the theoretical distribution.

Exercise D1.6

1 Mandy is testing a dice to see whether it is fair. She rolls the dice 600 times, and gets the results shown in the table.

Score	1	2	3	4	5	6
Frequency	98	103	107	105	97	90

   a Work out the theoretical probability of each score.
   b Use the table to estimate the experimental probability of each score to 3 dp.
   c Explain whether you think the dice is fair.

2 Mandy now tests a four-sided spinner marked 1–4. She spins it 500 times, and gets these results.

Score	1	2	3	4
Frequency	116	109	96	179

   a Work out the theoretical and experimental frequencies for each score.
   b Explain whether you think the spinner is fair.

3 A program picks random numbers between 1 and 8. Here is a sample of 80 numbers picked by the program.

5	5	3	8	6	7	5	6	3	4	4	8	6	8	5	3
5	4	3	8	4	4	6	4	3	8	3	2	4	2	8	7
6	1	5	4	1	8	5	5	7	6	1	5	6	1	2	1
3	1	1	4	8	3	7	3	8	6	7	4	6	6	8	3
2	4	2	2	2	3	4	3	3	1	2	8	4	1	8	1

a Use the data to estimate the probability of each digit being chosen.
b Explain whether you think the computer is picking the digits randomly.

4 The two spinners shown are spun, and the scores are added. The results for 100 trials are shown in the table.

Score	2	3	4	5	6	7	8	9
Frequency	5	11	14	23	17	16	10	4

   a Use a sample space diagram to find the theoretical probability of each result.
   b Use the table to estimate the experimental probability of each result.
   c Explain whether you think the spinners are fair.

5 Another experiment uses two spinners like those in question **4** but now the scores are multiplied. Here are the results for 1000 trials:

Result	1	2	3	4	5	6	8	9	10	12	15	16	20
Frequency	49	105	110	156	46	102	99	44	38	110	45	45	51

By comparing the theoretical and experimental probabilities for each result, explain whether or not you think that the spinners used in this experiment were fair.

This spread will show you how to:

▶▶ Know that a recurring decimal is a fraction.

▶▶ Use division to convert a fraction to a decimal.

▶▶ Order fractions by writing them with a common denominator or by converting them to decimals.

KEYWORDS

Recurring decimal

Terminating decimal

Denominator

A fraction describes a proportion of a whole.

In this pie chart:

$\frac{1}{8}$ is orange

$\frac{3}{8}$ is green

$\frac{4}{8}$ is red

▶ A fraction is the result of dividing one whole number by another. You divide to express a fraction as a decimal:

$\frac{3}{5} = 3 \div 5 = 0.6$

$2\frac{1}{5} = \frac{11}{5} = 11 \div 5 = 2.2$

You can express a number as a fraction or decimal of another number.

example

What fraction of:

a 81 is 36

b 2.4 m is 150 cm?

a As a fraction $\frac{36}{81} = \frac{4}{9}$

You can say $\frac{4}{9}$ of 81 is 36.

As a decimal $\frac{4}{9} = 4 \div 9 = 0.444 \ldots$

0.444 ... is a recurring decimal.

b Make the units the same: 2.4 m = 240 cm

As a fraction $\frac{150}{240} = \frac{5}{8}$.

You can say $\frac{5}{8}$ of 2.4 m is 150 cm.

As a decimal $\frac{5}{8} = 5 \div 8 = 0.625$

0.625 is a terminating decimal.

▶ Any decimal can be written as a fraction.

$1.45 = 1\frac{45}{100} = 1\frac{9}{20}$

You can order fractions using equivalent fractions or decimals.

example

Which is larger, $\frac{5}{8}$ or $\frac{7}{12}$?

Using fractions:

$\frac{5}{8} = \frac{15}{24}$ and $\frac{7}{12} = \frac{14}{24}$

$\frac{15}{24} > \frac{14}{24}$ so $\frac{5}{8} > \frac{7}{12}$

or

Using decimals:

$\frac{5}{8} = 0.625$ and $\frac{7}{12} = 0.58333 \ldots$

$0.625 > 0.58333 \ldots$ so $\frac{5}{8} > \frac{7}{12}$

Exercise N2.1

1 a Convert these fractions into decimals. You should be able to do some of the questions mentally.

i $\frac{3}{20}$ **ii** $\frac{44}{200}$ **iii** $\frac{13}{5}$ **iv** $\frac{14}{18}$ **v** $\frac{12}{60}$ **vi** $\frac{30}{16}$

b Convert these numbers to fractions in their simplest form.

i 0.68 **ii** 0.275 **iii** 3.6 **iv** 1.48

2 These diagrams show the number of students in two schools who have different kinds of pets. What proportion of the students in each school have a horse?

a Bar-de-Chart School

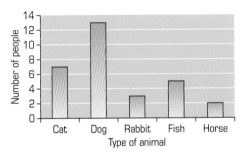

b Pie Technology School

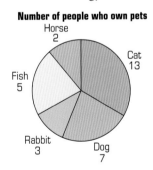

Number of people who own pets

c

Table Top Academy	
Types of pet	Number of students
Cat	45
Dog	30
Rabbit	16
Fish	20
Horse	8

3 What fraction of:

a 160 is 45

b 4 hours is 85 mins

c 3 km is 1240 metres

d 1 year is March

e 220 is 300?

Leave your answers in their simplest form.

4 For each pair of numbers insert >, < or = between them to make a true expression. Show your working out.

a $\frac{3}{8}$ 0.38

b 0.7   0.69

c $\frac{3}{7}$ $\frac{5}{11}$

d $\frac{3}{4}$ of a km $\frac{8}{11}$ of a km

e $\frac{25}{7}$ $3\frac{2}{5}$

f $2\frac{7}{12}$ $\frac{30}{11}$

g $\frac{21}{35}$ 0.585

h 2.42 $\frac{37}{15}$

5 a Put these fractions in order of size:

$\frac{7}{30}$ $\frac{26}{105}$ $\frac{4232}{16\,984}$ $\frac{123}{497}$ $\frac{576}{2311}$

b Find a fraction that is exactly halfway between $\frac{3}{7}$ and $\frac{10}{21}$.

6 Investigation

Investigate which fractions are represented by recurring decimals, and which give terminating decimals, for example:

$\frac{1}{3} = 0.333\,333\,333\,\dots$

$\frac{1}{5} = 0.2$

$\frac{1}{7} = 0.142\,857\,14\,\dots$

7 In a survey, Jacob asked over 24 000 people to pick their favourite colour. $\frac{43}{101}$ of the men preferred blue. $\frac{3}{7}$ of women preferred blue. Jacob claims that his results show more women prefer blue than men. Do you agree with his conclusion? Explain your reasons.

This spread will show you how to:

▶▶ Add and subtract fractions by writing them with a common denominator.

▶▶ Understand addition and subtraction of fractions.

▶▶ Use the laws of arithmetic and inverse operations.

KEYWORDS

Numerator
Denominator
Equivalent fraction
Lowest common multiple

You can add or subtract fractions with the same denominators.

$$\frac{2}{9} + \frac{3}{9} = \frac{5}{9}$$

$$\frac{2}{9} + \frac{3}{9} = \frac{5}{9}$$

▶ You can use equivalent fractions to add or subtract fractions with different denominators.

You find equivalent fractions with a common denominator.

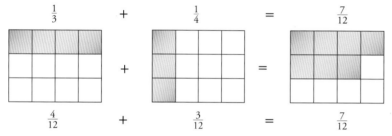

$$\frac{1}{3} + \frac{1}{4} = \frac{7}{12}$$

$$\frac{4}{12} + \frac{3}{12} = \frac{7}{12}$$

example

Work out $1\frac{5}{6} - \frac{8}{9}$.

The **lowest common multiple** of 6 and 9 is 18.
You use a common denominator of 18.

$$1\frac{5}{6} - \frac{8}{9} = \frac{11}{6} - \frac{8}{9}$$

$$= \frac{33}{18} - \frac{16}{18} = \frac{33 - 16}{18}$$

$$= \frac{17}{18}$$

$$\overset{\times\,\mathbf{3}}{1\frac{5}{6} = \frac{11}{6} = \underset{\times\,\mathbf{3}}{\frac{33}{18}}} \qquad \overset{\times\,\mathbf{2}}{\frac{8}{9} = \underset{\times\,\mathbf{2}}{\frac{16}{18}}}$$

You can change the fractions to decimals and then add or subtract.

example

Find the total of $\frac{3}{5} + 0.45 + \frac{1}{4}$.

Using equivalent fractions:

$$\frac{3}{5} = \frac{60}{100} \qquad \frac{1}{4} = \frac{25}{100} \qquad 0.45 = \frac{45}{100}$$

$$\frac{60 + 25 + 45}{100} = \frac{130}{100} = 1\frac{3}{10}$$

Using decimals:

$$\frac{3}{5} = 3 \div 5 = 0.6$$

$$\frac{1}{4} = 1 \div 4 = 0.25$$

$$0.6 + 0.25 + 0.45 = 1.3$$

Exercise N2.2

1 Work these out. Give your answers as mixed numbers where appropriate.

a $\frac{2}{5} + \frac{1}{5}$ b $\frac{3}{7} + \frac{3}{7} + \frac{4}{7}$

c $2\frac{1}{4} - \frac{3}{4}$ d $3\frac{1}{5} - \frac{4}{5}$

e $2\frac{4}{7} - 1\frac{6}{7}$ f $1\frac{5}{6} + 2\frac{5}{6}$

2 Use the fraction wall to make up five additions and subtractions using fractions.
For example: $\frac{2}{8} + \frac{1}{4} = \frac{1}{2}$
$$\frac{1}{2} - \frac{1}{3} = \frac{1}{6}$$

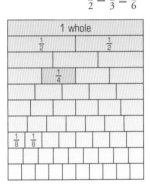

3 **Puzzle**
The Egyptians used to write fractions as the sums of unit fractions.
For example: $\frac{5}{6} = \frac{1}{2} + \frac{1}{3}$

a How would you write these Egyptian fractions?
 i $\frac{1}{3} + \frac{1}{4}$
 ii $\frac{1}{2} + \frac{1}{5} + \frac{1}{8}$
 iii $\frac{1}{4} + \frac{1}{15}$
 iv $\frac{1}{3} + \frac{1}{4} + \frac{1}{5}$

b How would the Egyptians write these fractions?
 i $\frac{5}{12}$ ii $\frac{8}{15}$
 iii $\frac{11}{12}$ iv $\frac{17}{20}$
 Explain your method for writing a fraction as an Egyptian fraction.

c Find three ways of writing 1 as the sum of unit fractions.

4 Work these out. Give your answers as mixed numbers where appropriate.

a $\frac{2}{7} + \frac{6}{14}$ b $3\frac{4}{7} - \frac{12}{14}$

c $\frac{2}{5} + \frac{3}{10}$ d $\frac{11}{3} - \frac{4}{5}$

e $\frac{5}{12} + \frac{4}{9}$ f $\frac{7}{8} - \frac{5}{12}$

g $\frac{17}{6} + 2\frac{1}{4}$ h $2\frac{5}{16} - \frac{11}{12}$

5 Solve these problems. Give your answers as mixed numbers where appropriate.

a Rufus mixed $1\frac{5}{8}$ litres of red paint with $2\frac{1}{5}$ litres of yellow paint to make orange paint. How many litres of orange paint did he make?

b A Banana Surprise pudding for 8 people requires 4 kg of bananas. Tim buys $2\frac{4}{7}$ kg of bananas. He already has 7 bananas each weighing $\frac{1}{5}$ kg. Does he have enough bananas to make the Banana Surprise?

c A plank of wood is 17.4 m long. Angus saws off the following pieces:

> 2 pieces @ $1\frac{1}{5}$ metres
> 1 piece @ 3.65 metres
> 3 pieces @ $2\frac{3}{8}$ metres

What is the new length of the plank?

6 **Investigation**
 ▶ Choose two fractions:
 For example: $\frac{3}{5}$ and $\frac{2}{7}$.
 ▶ Add the fractions:
 $\frac{3}{5} + \frac{2}{7} = \frac{31}{35}$.
 ▶ Swap the numerators and add the new fractions:
 $\frac{2}{5} + \frac{3}{7} = \frac{29}{35}$.
Investigate swapping the numerators with different pairs of fractions.
See if the sum of the fractions increases, decreases or stays the same.
Write down what you notice.

This spread will show you how to:
- ▶▶ Calculate fractions of quantities.
- ▶▶ Multiply and divide an integer by a fraction.
- ▶▶ Consolidate and extend mental methods of calculation.

You can multiply a unit fraction by an integer using a number line:

$$\tfrac{1}{5} \times 3 = 3 \times \tfrac{1}{5} = \tfrac{3}{5}$$

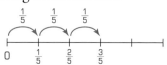

▶ **You can use unit fractions to multiply any fraction by an integer.**

$$\tfrac{3}{5} \times 2 = (\tfrac{1}{5} \times 3) \times 2$$
$$= \tfrac{1}{5} \times 6 = \tfrac{6}{5}$$

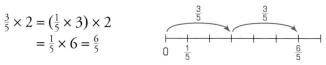

▶ **Multiplying by $\tfrac{1}{5}$ is the same as dividing by 5.**

You can cancel to simplify the product of a fraction and an integer.
This can be done at the end of the calculation ...   ... or during the calculation.

$$\tfrac{3}{8} \times 20 = (\tfrac{1}{8} \times 3) \times 20 = \tfrac{3 \times 20}{8} = \tfrac{60}{8} = \tfrac{15}{2} = 7\tfrac{1}{2}$$
(÷4 shown above and below the fraction)

$$\frac{3 \times \cancel{20}^{5}}{\cancel{8}_{2}} = \frac{3 \times 5}{2} = \frac{15}{2} = 7\tfrac{1}{2}$$

You can calculate a fraction of an amount in different ways:

Use the relationship between fractions and division:	Multiply the fraction by the amount:	Use an equivalent decimal:
$3 \times \tfrac{1}{5}$ of 128 cm	$\tfrac{3}{5}$ of 128 cm	$\tfrac{3}{5} \times 128$ cm
$= 3 \times (\tfrac{1}{5}$ of 128 cm)	$= \tfrac{3}{5} \times 128$ cm	$= (3 \div 5) \times 128$ cm
$= 3 \times (128 \div 5)$ cm	$= \tfrac{1}{5} \times 3 \times 128$ cm	$= 0.6 \times 128$ cm
$= 76.8$ cm	$= 76\tfrac{4}{5}$ cm	$= 76.8$ cm

▶ **You can use unit fractions to divide an integer by a fraction.**

▶ **Dividing by $\tfrac{1}{5}$ is the same as multiplying by 5.**

$1 \div \tfrac{1}{5} = 1 \times 5 = 5$

How many fifths are
there in 1 whole?

$2 \div \tfrac{1}{5} = 2 \times 5 = 10$

How many fifths are
there in 2 wholes?

$4 \div \tfrac{2}{3}$
$4 \div \tfrac{1}{3} = 4 \times 3 = 12$
so $4 \div \tfrac{2}{3} = 6$

How many $\tfrac{2}{3}$ are
there in 4 wholes?

Exercise N2.3

1 Copy and complete these fraction times tables:

a $\frac{1}{7} \times 1 = \frac{1}{7}$ **b** $\frac{2}{7} \times 1 = \frac{2}{7}$
$\frac{1}{7} \times 2 = \frac{2}{7}$ $\frac{2}{7} \times 2 =$
$\frac{1}{7} \times 3 =$ $\frac{2}{7} \times 3 =$

c $\frac{3}{7} \times 1 =$
$\frac{3}{7} \times 2 =$
$\frac{3}{7} \times 3 =$

2 Use number lines to calculate:

a $\frac{2}{3} \times 5$

b $\frac{3}{5} \times 4$

c $5 \times \frac{4}{9}$

3 Calculate these, giving your answer as a mixed number where appropriate:

a $7 \times \frac{3}{8}$ **b** $12 \times \frac{5}{6}$

c $\frac{3}{7} \times 21$ **d** $\frac{7}{16} \times 24$

e $20 \times \frac{15}{8}$ **f** $28 \times \frac{11}{21}$

g $\frac{7}{9} \times 24$ **h** $2\frac{3}{8} \times 6$

4 Calculate these fractions of amounts using the most appropriate method. Give your answers to 2 dp:

a $\frac{3}{5}$ of 19

b $\frac{13}{15}$ of 255 cm

c $\frac{2}{13}$ of £65

d $\frac{7}{15} \times 40$ m

e $\frac{4}{9}$ of 125 g

f $\frac{8}{5} \times 85$ litres

g $3\frac{3}{7}$ of 182 sheep

h $1\frac{3}{8}$ of 27 km.

5 Puzzle

True or false?
$\frac{3}{4}$ of $\frac{4}{5}$ of $\frac{2}{3}$ of £120 $> \frac{2}{3}$ of $\frac{3}{4}$ of $\frac{4}{5}$ of £120?
Explain your answer.

6 Alison's working day is 9 hours long. She spends $\frac{7}{12}$ of her day working at the computer. Of her remaining time she spends $\frac{7}{9}$ answering the phone, and the rest at lunch.

a How long is Alison's lunch break?

b What fraction of her day is spent on her lunch break?

7 Calculate:

a $3 \div \frac{1}{4}$ **b** $5 \div \frac{1}{3}$

c $4 \div \frac{2}{5}$ **d** $12 \div \frac{3}{7}$

e $36 \div 2\frac{2}{5}$

8 Investigation

a What is the smallest integer that can be multiplied by $\frac{2}{9}$ so that the result is a mixed number whose fractional part is also $\frac{2}{9}$? Here is an example for $\frac{2}{5}$:

$$\frac{2}{5} \times 11 = \frac{22}{5} = 4\frac{2}{5}$$

b Investigate for any starting fraction. Explain your answer.

9 a Toby has £2340 to invest in the stock market. He uses $\frac{4}{7}$ of his money to buy shares in Marks and Spencer.
He spends $\frac{7}{12}$ of the remaining money on 117 shares in Debenhams.
How much does each Debenhams share cost?

b Sarah-Jane has just won some money on the National Lottery. She uses $\frac{3}{5}$ of her winnings to buy a new car and $\frac{1}{16}$ of her winnings to pay for the insurance. With the remainder of her winnings she spends $\frac{1}{3}$ on a holiday, $\frac{1}{9}$ on new furniture and invests the remaining £2250 as a deposit on a new home.
How much money did she spend on each item?

This spread will show you how to:
▶▶ Interpret percentage as the operator 'so many hundredths of'.
▶▶ Calculate percentages.
▶▶ Consolidate and extend mental methods of calculation.
▶▶ Recall known facts and use them to derive unknown facts.

KEYWORDS
Percentage
Equivalent
Unitary method

▶ A percentage is a fraction out of 100.
   12% means 12 parts per 100 (or 12 parts in every 100).

You can calculate the percentage of an amount using a range of mental, written and calculator methods.

Remember:
▶ Finding 50% is the same as finding $\frac{1}{2}$ or dividing by 2.
▶ Finding 10% is the same as finding $\frac{1}{10}$ or dividing by 10.
▶ Finding 1% is the same as finding $\frac{1}{100}$ or dividing by 100.

Mental methods

Using 10%

To find 60% of 80:
50% of 80 = 40
10% of 80 = 8
60% of 80 = 48

Unitary method

To find 12% of £300:
1% of £300 = £300 ÷ 100 = £3
12% of £300 = 12 × £3 = £36

Written methods

Using an equivalent fraction

To find 23% of 30:
$23\% \text{ of } 30 = \frac{23}{100} \times 30$
$= \frac{1}{100} \times 23 \times 30$
$= \frac{23 \times 30^3}{100^{10}} = \frac{69}{10}$
$= 6.9$

Using an equivalent decimal

To find 23% of 30:
$23\% \text{ of } 30 = \frac{23}{100} \times 30$
$= (23 \div 100) \times 30$
$= 0.23 \times 30$
$= 6.9$

Calculator methods

The best way to calculate is to estimate first then use a decimal.

To find 37% of 82:
$37\% \text{ of } 82 = 0.37 \times 82$ Estimate: $0.4 \times 80 = 32$

Calculate: [0] [.] [3] [7] [×] [8] [2] [=]

The display should show 30.34

Exercise N2.4

1 Calculate these, using a mental or written method as appropriate:

  a 10% of 340 m   **b** 60% of 150 kg

  c 15% of £27   **d** 125% of 84 cm

  e 11% of 65 litres   **f** 35% of 720 tins

  g 95% of £240   **h** 115% of £27

2 **Puzzle**

Match the percentages, amounts and answers to make six correct statements.

Percentages	Amounts	Answer
18%	£38	£8.19
13%	£35	£7.74
47%	£43	£11.90
34%	£63	£10.81
29%	£27	£11.02
43%	£23	£11.61

3 **a** An alloy contains 35% silver, 43% nickel and 22% iron. How much nickel is there in 245 kg of the alloy?

  b The Recommended Daily Allowance (RDA) of iron is 14 mg. A bowl of cereal provides 37% of the RDA of iron. How much iron is there in a bowl of cereal?

  c A suit is designed from a material containing 14% cotton. If the suit weighs 485 grams, what is the weight of cotton in the suit?

4 An examination is marked out of 180. The highest mark is 95%. The lowest mark is 15%. What is the difference between the highest and lowest marks?

5 Three women ran for eight minutes around a track.

    Rachel ran $\frac{3}{7}$ of 1500 m

    Phoebe ran 65% of 1 km

    Monica ran 161% of 400 m

  a How far did each woman run?

  b How far ahead of the last runner was the first runner?

6 Copy and complete this table of the distance travelled to school each day by students from Green School.

Distance	Number of students	Percentage
<1 km		21.0
1 to <3 km		33.0
3 to <5 km		14.6
5 to <10 km		15.0
10 km & over		16.4
Total	1500	100.0

7 Calculate these, giving your answer in metres (to 2 dp where appropriate).

  a 43% of 85 m   **b** 156% of 320 cm

  c 12% of 48 m   **d** 6% of 0.84 km

  e 11.25% of 12 400 mm

  f $33\frac{1}{3}$% of 4.5 m

8 **Puzzle**

  a Liam was writing up his report on a survey to find the most popular flavour of crisps in Year 8 at his school. He wrote, '20% of Year 8 (52 pupils) preferred Cheese and Onion crisps!' How many pupils are there in Year 8 at Liam's school?

  b 45% of the population of Keswick own bicycles. There are 3060 bicycles in Keswick. What is the population of Keswick?

Percentage increase and decrease

This spread will show you how to:
▶▶ Interpret percentage as the operator 'so many hundredths of'.
▶▶ Calculate percentages and find the outcome of a given percentage increase or decrease.

KEYWORDS
Percentage   Decrease
Increase
Value added tax (VAT)
Service charge

In real life, lots of things are increased or reduced by a percentage.

SALE all prices reduced by 30%

WORKERS DEMAND A 6% INCREASE

Shares have fallen by a record 35% over the past 6 months

You can add on a percentage increase or subtract a percentage decrease at the end of the calculation.

example

A book costing £18 is reduced in a sale by 22%.
What is the sale price?

....

Original price = £18   Reduced means decrease.   Percentage decrease = 22%

Reduction = 22% of the original price = 22% of £18
$$= £3.96$$

Sale price   = original price − reduction = £18 − £3.96
$$= £14.04$$

Alternatively you can calculate the percentage increase or decrease in a single calculation.

example

a A book costing £18 is reduced in a sale by 22%. What is the sale price?

b A computer costing £540 is increased in price by 15%.
What is the new price of the computer?

....

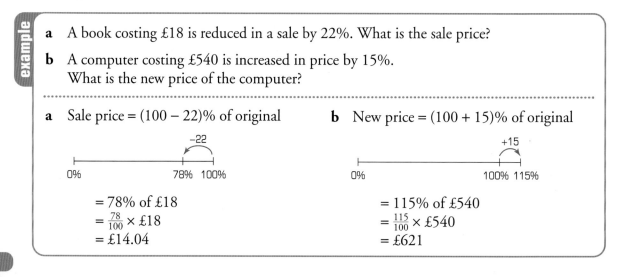

a Sale price = (100 − 22)% of original

$$= 78\% \text{ of } £18$$
$$= \frac{78}{100} \times £18$$
$$= £14.04$$

b New price = (100 + 15)% of original

$$= 115\% \text{ of } £540$$
$$= \frac{115}{100} \times £540$$
$$= £621$$

Exercise N2.5

1 Calculate these, using the most appropriate method:

  a 10% of £715   **b** 30% of 250 cakes

  c 120% of 45 m   **d** 12% of 85 kg

  e 99% of 3400 m   **f** 105% of £2850

  g 73% of €4650   **h** 37% of 65 min

2 **a** A rectangular piece of paper measures 24 cm by 18 cm.
    35% of the paper is painted blue.
    What area of paper is not painted blue?

  b Jo invests £8364.00.
    Three years later her money is worth 118% of its original value.
    What is the new value of her savings?

  c Amy bought a set of chairs for £1160.
    A year later she sells the chairs for 82% of the price she paid.
    For how much did she sell the chairs?

3 **a** A meal costs £28 plus a 12% service charge. How much is the bill?

  b During a '21st Year Anniversary' sale all items are reduced by 21%.
    What is the new price of a suit costing £385 before the sale?

  c Gill used to weigh 72 kg. After going on a diet she has reduced her weight by 24%. What is her new weight?

  d A shirt is originally priced at £19, but during a sale it is reduced by 35%.
    How much does the shirt cost now?

  e Dermott buys 26 paving stones at £3.80 each and 3 bags of cement at £5.25 a bag.
    He is charged VAT at 17.5% on all items. What is his final bill (to the nearest penny)?

  f The population of Manchester in 1881 was about 1 148 000.
    During the next 120 years the population increased by 220%.
    What was the population of Manchester in 2001?

4 When measuring distances, Eddie knows that his instruments have an error of up to 2%. If he measures the length of a room as 640 cm, what is the:

  a greatest possible length of the room
  b least possible length of the room?

5 **Puzzle**
Hector says, 'If you change something by 1% then it doesn't really make any difference – you've still got most of what you started with!

> For example: 99% of £500 = £495 and £495 is about £500 to the nearest £10.'

Fiona disagrees with Hector.
Who is correct?
Explain and justify your answer.

6 Susan can buy a Games Console for one cash payment of £199, or pay a deposit of 22% and then twelve equal monthly payments of £14.

Which is the better option?
Explain and justify your answer.

7 **Investigation**
In February, a camera is on sale in PhotoWorld for £289.
In March, the manager reduces the price by 15%.
In April, she reduces the new price by a further 15%.
She makes a banner saying that the camera has been reduced by 30% since February.

Is the banner telling the truth?
Explain and justify your answer.

Investigate the effect of combining other percentage reductions.

This spread will show you how to:

▶▶ Express one given number as a percentage of another.

▶▶ Use the equivalence of fractions, decimals and percentages to compare simple proportions.

▶▶ Recall known facts and use them to derive unknown facts.

KEYWORDS

Percentage

Compare

Proportion

A proportion is a part of the whole.
You can use percentages, fractions and decimals to describe proportions.

The results of this survey illustrate various proportions:

A typical Year 8 student spends:

▶ $\frac{8}{24}$ or $\frac{1}{3}$ of the day sleeping
▶ 25% of the 24 hours working
▶ 5 out of the 24 hours watching TV
▶ 0.125 of the day doing other things.

How Year 8 students spend a typical day

Other 3hrs

Sleeping 8hrs

TV 5hrs

Eating 2hrs

Working 6hrs

▶ To convert a fraction or decimal to a percentage you multiply by 100.

$0.125 \times 100\% = 12.5\%$

$\frac{5}{24} \times 100\% = 5 \div 24 \times 100\% = 20.8\%$ (1 dp)

example

Express 24 hours as a proportion of 168 hours.

Proportion $= \frac{24}{168} = \frac{1}{7}$

$= 24 \div 168 = 0.143$ (3 dp)

$= (24 \div 168) \times 100\% = 14.3\%$ (1 dp)

So, 1 day (24 hours) is ...
... $\frac{1}{7}$ of 1 week (168 hours)
... 0.143 of 1 week (168 hours)
... 14.3% of 1 week (168 hours)

You can use proportions to make simple comparisons.

example

The information for three types of cereal states that:

▶ Choco Pops are 9.6% fat
▶ Crisp Flakes contain 2.7 g fat per 30 g serving
▶ Weetybiscs contain 71 g of fat in each 750 g packet.

Which cereal contains the least fat?

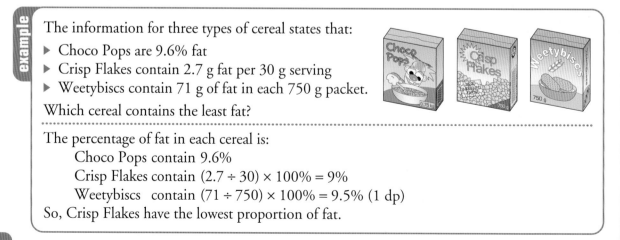

The percentage of fat in each cereal is:

Choco Pops contain 9.6%

Crisp Flakes contain $(2.7 \div 30) \times 100\% = 9\%$

Weetybiscs contain $(71 \div 750) \times 100\% = 9.5\%$ (1 dp)

So, Crisp Flakes have the lowest proportion of fat.

Exercise N2.6

1. Copy and complete these equivalent fractions, decimals and percentages. Work them out without using a calculator.

Fraction	Decimal	Percentage
$\frac{2}{5}$		
	0.65	
		48%
$\frac{7}{8}$		
	1.28	

2. **a** 1 kilogram of beef contains 250 grams of fat. What percentage is not fat?
   b To pass his maths exam, Jochim needs to score 80 marks out of 200. What percentage does he need to pass?
   c David earns £25 000 a year. He pays £6000 in tax. What percentage of his earnings does he pay in tax?

3. Two photocopiers are compared to see which makes the most errors.
   The X 40E makes 13 errors in 250 copies.
   The Z H 2 makes 9 errors in 180 copies.
   Which photocopier is more reliable?
   Explain your answer.

4. Over a season, Dwight Cole scored 3 goals every 7 games, Ryan Beckham scored a goal in 42.3% of his games, Eric Rivaldo scored 17 goals in 42 games.
   Who was the better striker?
   Explain and justify your answer.

5. Jim has invested £230 in Berkleys Bank for 1 year. At the end of the year he receives £11 in interest.
   Ingrid has invested £180 in Madbury Supermarket bank for 1 year. At the end of the year she receives £9 in interest.
   Who has had the better rate of interest, Ingrid or Jim? Explain your answer.

6. Mylene asked 120 pupils in Year 8 which subject they liked best.

Subject	Boys	Girls
Science	10	14
English	7	15
Maths	13	13
Art	2	8
PE	10	2
Languages	1	9
History	3	7
Geography	4	2
Total	50	70

   a Which subject was chosen by 8% of boys?
   b What proportion of the girls chose Languages?
   c Ilyas said that Maths was equally popular with boys and girls.
   Catherine said that Science was equally popular with boys and girls.
   Explain how Ilyas and Catherine have arrived at their different answers.

7. **Investigation**
   Three different ores are blended together to make an iron and nickel alloy which is exactly 65% iron and 10% nickel.

Ore	Iron content	Nickel content
A	80%	12%
B	70%	8%
C	60%	10%

   How would you mix the three ores to give the correct iron and nickel alloy?

You should know how to ...

1 Use the equivalence of fractions, decimals and percentages to compare proportions.

2 Calculate percentages and find the outcome of a given percentage increase or decrease.

3 Add and subtract fractions by writing them with a common denominator.

Check out

1 a What is 35 as a proportion of 120?
Write your answer as
i a fraction
ii a percentage.

b In a quiz show Bill Brainy answered 12 of his 17 questions correctly.
Steve Stupid got 13 questions wrong out of 48.
Who got the highest proportion of correct answers?

c In a survey of favourite animals, $\frac{3}{7}$ of boys and 41% of girls preferred cats.
Who preferred cats the most, boys or girls?

2 a Calculate 27% of 456 m.

b A piece of elastic is 88 cm long.
When stretched its length increases by 82%.
What is its new length?

c A new car costs £12 500.
It reduces in value by 15% a year.
What is its value after a year?

3 a Calculate $\frac{3}{7} + \frac{6}{5}$.

b These three fractions add to make $\frac{1}{2}$:
$$\frac{1}{5} + \frac{1}{4} + \frac{1}{20}$$
Find three other fractions with different denominators which add to make $\frac{1}{2}$.

Expressions and formulae

This unit will show you how to:

▶▶ Begin to distinguish the different roles played by letter symbols in equations, formulae and functions.

▶▶ Know that algebraic operations follow the same conventions and order as arithmetic operations.

▶▶ Use index notation for small positive integer powers.

▶▶ Simplify or transform linear expressions by collecting like terms.

▶▶ Multiply a single term over a bracket.

▶▶ Use formulae from mathematics and other subjects.

▶▶ Substitute integers into simple formulae and positive integers into expressions involving small powers.

▶▶ Solve more demanding problems and investigate in the context of algebra.

▶▶ Represent problems in algebraic form using correct notation and appropriate diagrams.

Computer language is based on algebra.

Before you start

You should know how to ...

1 Use letter symbols to represent unknown numbers.

2 Recognise and use basic algebraic conventions.

▶ $3n$ means $n + n + n$

3 Add and subtract positive and negative numbers.

For example:

▶ $^-3 + \,^-2 = \,^-1$ ▶ $^-3 + \,^-2 = \,^-5$ ▶ $3 + \,^-2 = 1$

Check in

1 Write these sentences using algebra:
   a 6 less than x then halve
   b 5 more than y then double.

2 Simplify these expressions:
   a $x + y + x + y$ **b** $x - 3x + 6x$

3 Calculate:
   a $5 + \,^-7$ **b** $11 - \,^-9$
   c $^-7 + 3$ **d** $^-12 - \,^-15$

This spread will show you how to:

▶▶ Know that algebraic operations follow the same conventions and order as arithmetic operations.

▶▶ Substitute integers into simple linear formulae.

KEYWORDS

Expression   Equivalent
Equation   Substitute

An algebraic expression is a collection of terms.

$3x + y - 2$ has three terms: $3x$, y and 2.

You can find the value of an expression when you know the value of the letters used.
First you need to know the rules used in algebraic expressions:

▶ Do not use the multiplication or division sign.

$4 \times y$ is $4y$, and $(3x - 7) \div 6$ is $\dfrac{3x - 7}{6}$

▶ Work out brackets first, then multiply and divide, then add and subtract.

Order these expressions from smallest to largest when $x = 7$ and $y = {}^-2$.

a $4x - 2y$ **b** $2xy - 5$ **c** $3(x - y)$ **d** $\dfrac{5x - 3y}{2}$

Substitute the values of x and y into the expressions.

a $4x - 2y$
$= 4 \times 7 - 2 \times {}^-2$
$= 28 + 4$
$= 32$

b $2xy - 5$
$= 2 \times 7 \times {}^-2 - 5$
$= {}^-28 - 5$
$= {}^-33$

c $3(x - y)$
$= 3 \times (7 - {}^-2)$
$= 3 \times 9$
$= 27$

d $\dfrac{5x - 3y}{2}$
$= (5x - 3y) \div 2$
$= (5 \times 7 - 3 \times {}^-2) \div 2$
$= (35 + 6) \div 2$
$= 20.5$

In order the expressions are:
$2xy - 5$, $\dfrac{5x - 3y}{2}$, $3(x - y)$, $4x - 2y$

You can only use the equals sign when all the sides are equal.

An equation has a particular value:
$2x - 5 = 9$, so $x = 7$.

You can write equivalent equations to $2x - 5 = 9$
using the terms $2x$, 5 and 9 in different ways:

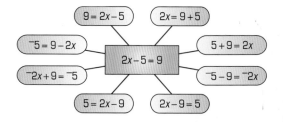

Exercise A2.1

1 Choose the correct answer to each question from this list:

a $14 - 3 \times 2$	⁻3
b $2 \times 7 - 10$	35
c $3 + 8 \times 4$	44
d $5 + 3 \times 6$	8
e $4(3 + 8)$	10
f $18 - 7 \times 3$	0
g $8 - 4(7 - 5)$	4
h $2 \times 7 - 2(5 - 3)$	23

Check you have only used each answer once.

2 All of these problems have the correct answer, but the working is wrong because the equals sign is not used correctly.
Rewrite each problem, setting out the working correctly.
The first one is done for you.

a $78 + 65 = 78 + 60 = 138 + 5 = 143$
$78 + 65 = 78 + 60 + 5 = 138 + 5 = 143$

b $187 + 49 = 187 + 50 = 237 - 1 = 236$
c $234 - 86 = 234 - 80 = 154 - 6 = 148$
d $89 \times 6 = 90 \times 6 = 540 - 6 = 534$

3 Find the value of these algebraic expressions when $p = 3$, $q = 5$ and $r = 6$.

a $\dfrac{(4p + r)}{2}$ **e** $6 - 2r$ **i** $r + pq$ **n** pqr

o $2(3q + 1)$ **q** $pq - 3r$ **t** $4 + 3p$ **u** $pr - 2q$

Rearrange your answers in order, smallest to largest.
What mathematical word do the letters spell?

4 You can rewrite the equation: $5x - 20 = 10$
as $10 = 5x - 20$.
Rewrite $5x - 20 = 10$ in six different ways.

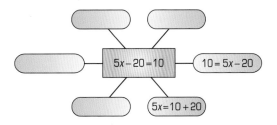

5 $2x + 7 = 11$ is the same as $11 - 2x = 7$
These equations make four pairs of equivalent equations. Find the four pairs.

a $12 + 3x = 21$ **b** $3x - 21 = 12$ **c** $21 + 3x = 12$
d $3x - 12 = 21$ **e** $21 - 3x = {}^-12$ **f** $21 = 12 + 3x$
g $^-12 = 21 - 3x$ **h** $12 = 3x + 21$

This spread will show you how to:
▶▶ Use index notation for small positive integer powers.
▶▶ Substitute positive integers into expressions involving small powers.

KEYWORDS
Index notation
To the power of n
Indices   Cubed

Using algebra notation you could write $p \times p$ as pp.

As in arithmetic, the convention is to write p^2.

Remember:
$3^2 = 3 \times 3$
$5 \times 5 = 5^2$

You write $p \times p \times p$ as p^3 (you say 'p cubed' or 'p to the power 3').

▶ **When you add a letter repeatedly you get a multiple of the letter.**
$n + n + n + n = 4n$

▶ **When you multiply a letter repeatedly you get a power of the letter.**
$p \times p \times p \times p = p^4$ (you say 'p to the power of 4')

You can simplify expressions involving powers.

For example: $p^3 \times p^2 = (p \times p \times p) \times (p \times p) = p^5$

When you multiply expressions in index notation, you add the indices.

You need to be careful when numbers and powers are combined:

$3n^2$ means $3 \times n \times n$.

---

example

Work out the value of these expressions when $p = 5$.

a p^3 **b** $3p^2$ **c** $(3 + p)^2$ **d** $3 + p^2$ **e** $\dfrac{2p^2}{10}$

....

a $p^3 = 5 \times 5 \times 5 = 125$

b $3p^2 = 3 \times 5 \times 5 = 75$

c $(3 + p)^2 = (3 + 5)^2 = 8^2 = 64$

d $3 + p^2 = 3 + 5 \times 5 = 3 + 25 = 28$

e $\dfrac{2p^2}{10} = \dfrac{2 \times 5 \times 5}{10} = \dfrac{50}{10} = 5$

Exercise A2.2

1 Write down these expressions in their simplest form:

a $2x + 3x - x$ **b** $p - 4x + 3p$

c $y \times y \times y \times y$ **d** $6x \times x \times y$

e $n^3 \times n^2 \times n$ **f** $2x \times 2x$

g $p^2 \times p^2 \times p^2$ **h** $x \times x \times x + y \times y$

2 Work out the value for each of these expressions when $e = 3$, $f = 4$ and $g = \frac{1}{2}$.

a $2f$ **b** f^2 **c** e^3 **d** $(3 + f)^2$

e $3 + f^2$ **f** $e^2 f$ **g** $5f^2$ **h** $e^2 + f^2$

i $(e + f)^2$ **j** e^4 **k** $4e^2$ **l** $e^3 + 3g$

m $e^2 fg$ **n** $(\frac{1}{2}e + g)^2$ **o** $7eg$ **p** $(f - e)^7$

q $(12g)^3$ **r** $e^4 - f^3$ **s** $(e + fg)^2$ **t** g^2

3 $2n$ and n^2 are different expressions.

  a Find a value for n to show that the two expressions have different values.

  b Find a value for n that gives the same value for each of the expressions.

  c Find a value for n where $2n$ is larger than n^2.

4 Here are 11 algebraic expressions:

  ▶ Match five pairs of expressions that have the same meaning.

  ▶ For the remaining expression, write a simplified expression which is equivalent.

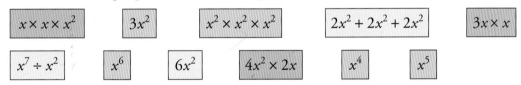

$x \times x \times x^2$	$3x^2$	$x^2 \times x^2 \times x^2$	$2x^2 + 2x^2 + 2x^2$	$3x \times x$
$x^7 \div x^2$	x^6	$6x^2$	$4x^2 \times 2x$	x^4	x^5

5 For each of these sets, find the expression, equation or formula which is the odd one out.

a $2x + 4 = 3x$; $4 + 2x = 3x$; $3x - 4 = 2x$; $4 - 3x = 2x$

b $p + q = r$; $q + p = r$; $r - p = q$; $p - r = q$

c $y = 3x - 4$; $^-4 + 3x = y$; $3x = y + 4$; $4 = 3x + y$

d $V = IR$; $RI = V$; $R = IV$; $R = \dfrac{V}{I}$

e $4p + 7 = 6p + 1$; $4p - 1 = 6p - 7$; $4p + 6p = 7 + 1$; $7 + 4p = 1 + 6p$

f $3 \times x^4$; $3 \times x \times x \times x \times x$; $3x + x + x + x$; $3x^6 \div x^2$

g $p^2 \div p^4$; $p \times p$; $p \times p \times p \div p$; $p \times p^3 \div p^2$

This spread will show you how to:

▶▶ Simplify or transform linear expressions by collecting like terms.

KEYWORDS

Collect like terms
Linear expressions
Simplify

There are four terms in this linear expression:

$5x$	$+3y$	$-x$	$+2$
x term	y term	x term	number

The two x terms can be simplified.
This is called collecting like terms.

$4x$ $+3y$ $+2$

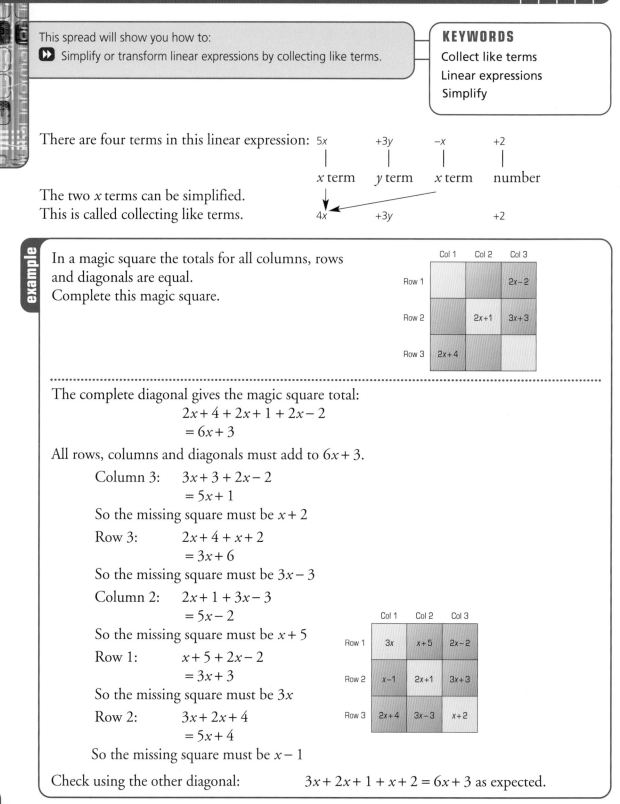

example

In a magic square the totals for all columns, rows and diagonals are equal.
Complete this magic square.

The complete diagonal gives the magic square total:
$$2x + 4 + 2x + 1 + 2x - 2$$
$$= 6x + 3$$

All rows, columns and diagonals must add to $6x + 3$.

Column 3: $3x + 3 + 2x - 2$
               $= 5x + 1$

So the missing square must be $x + 2$

Row 3: $2x + 4 + x + 2$
              $= 3x + 6$

So the missing square must be $3x - 3$

Column 2: $2x + 1 + 3x - 3$
               $= 5x - 2$

So the missing square must be $x + 5$

Row 1: $x + 5 + 2x - 2$
              $= 3x + 3$

So the missing square must be $3x$

Row 2: $3x + 2x + 4$
              $= 5x + 4$

So the missing square must be $x - 1$

Check using the other diagonal: $3x + 2x + 1 + x + 2 = 6x + 3$ as expected.

Exercise A2.3

1 In Hot Crosses the sums of the expressions in the horizontal row
and vertical column are the same.
Show that these are Hot Crosses.

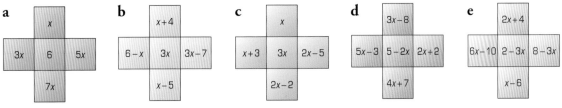

a

	x	
$3x$	6	$5x$
	$7x$	

b

	$x+4$	
$6-x$	$3x$	$3x-7$
	$x-5$	

c

	x	
$x+3$	$3x$	$2x-5$
	$2x-2$	

d

	$3x-8$	
$5x-3$	$5-2x$	$2x+2$
	$4x+7$	

e

	$2x+4$	
$6x-10$	$2-3x$	$8-3x$
	$x-6$	

2 In a magic square the sum of the expressions in each row, column and diagonal is the same.
Complete the magic squares:

a

		7
9	5	1

b

		$x+3$
$x+4$	x	$x-4$

c Show that this is a magic square:

$y+3$	$4y-1$	$4y+4$
$6y+3$	$3y+2$	1
$2y$	$2y+5$	$5y+1$

3 In these towers, add the adjacent cells together and put the answer in the cell below.

a

$x+3$	$2x-1$	$x+5$

b

$x-5$	$3x+1$	$2x+3$

c

$3x-7$	$2x+3$	$5x-2$

d

$4x+3$		$x+2$
	$6x+5$	

4 Complete the addition to find the tower totals.

a

$2a+3b$	$5a-2b$	$2a-b$

b

$3a-b$	$4b-2a$	$a+b$

c

$5b-4a$	$a+3b$	$2b-5a$

5 Here are five algebraic expressions:

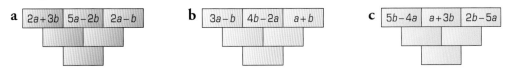

$$3x-2y \qquad 4x+5y \qquad 3x^2+4y^2 \qquad 3y-5x \qquad 2x^2-3y^2$$

Add or subtract two of the expressions to make each of these expressions.
The first one is done for you.

a $8y-x=(4x+5y)+(3y-5x)$

b $y-2x$ **c** $7x+3y$ **d** $5x^2+y^2$ **e** x^2+7y^2

This spread will show you how to:
⏩ Simplify or transform linear expressions by collecting like terms.
⏩ Multiply a single term over a bracket.

You can multiply out brackets in algebra using the grid method.

Remember: So:

$$76 \times 8 = 8 \times (70 + 6)$$ $$8(7t + 6)$$

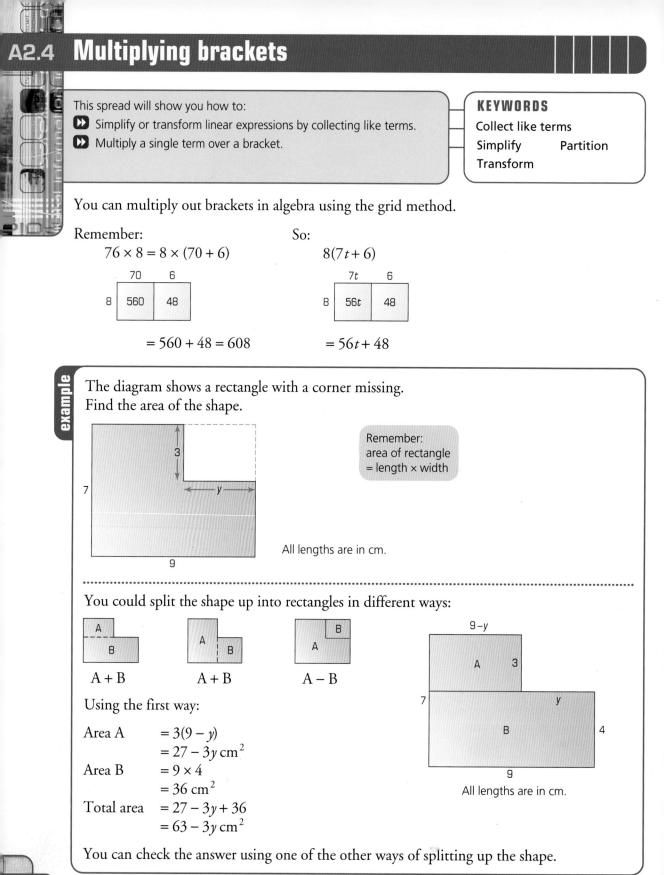

	70	6
8	560	48

	7t	6
8	56t	48

$$= 560 + 48 = 608$$ $$= 56t + 48$$

example

The diagram shows a rectangle with a corner missing.
Find the area of the shape.

Remember:
area of rectangle
= length × width

All lengths are in cm.

You could split the shape up into rectangles in different ways:

A + B   A + B   A − B

Using the first way:

Area A $= 3(9 - y)$
           $= 27 - 3y \text{ cm}^2$
Area B $= 9 \times 4$
           $= 36 \text{ cm}^2$
Total area $= 27 - 3y + 36$
           $= 63 - 3y \text{ cm}^2$

All lengths are in cm.

You can check the answer using one of the other ways of splitting up the shape.

Exercise A2.4

1 a Work out these expressions by partitioning.
The first one is done for you:

   i 3×24 $3 \times 24 = 3 \times 20 + 3 \times 4 = 60 + 12 = 72$

   ii 8×47 **iii** 7×34 **iv** 9×52

b Multiply out these expressions.
The first one is done for you:

   i $3(2t + 4)$ $3(2t + 4) = 3 \times 2t + 3 \times 4 = 6t + 12$

   ii $8(4t + 7)$ **iii** $7(3t + 4)$ **iv** $9(5t + 2)$

2 Simplify these expressions and find the odd one out in each set.
The first one is done for you.

  a $8(r + 3s)$; $3(2r + 8s)$; $4(2r + 6s)$

     $8(r + 3s) = 8r + 24s$; $3(2r + 8s) = 6r + 24s$; $4(2r + 6s) = 8r + 24s$

     so $3(2r + 8s)$ is the odd one out.

  b $4(2a + 3b)$; $2(4a + 6b)$; $3(3a + 4b)$
  c $6(3x - 4y)$; $4(4x - 6y)$; $3(6x - 8y)$
  d $a(6a - 8b)$; $3a(2a - 3b)$; $2a(3a - 4b)$
  e $5(3u - 2v) + 3v$; $8(2u - v) + v$; $3(4u - 2v) + 3u - v$
  f $5p + 3(4p - 2q)$; $3q + 5(3p - 2q)$; $3(5p - 3q) + 2q$

3 Simplify each of these expressions by multiplying out the brackets and collecting like terms.

  a $3(x + 7y) + 2(3x + y)$ **b** $4(2r - 3s) + 3(r + 2s)$
  c $5(3j - 2k) - 2(5j + 3k)$ **d** $4(7p + 3q) - 3(2p - 5q)$
  e $x(1 + 2y) + y(5x - 2)$ **f** $6(y - 3x) - x(2 + y)$

4 Copy these diagrams.

  a Add three different ways
of writing $36x - 24$

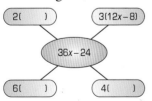

  b Add different ways of
representing $18pq - 12p$

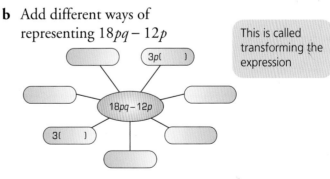

This is called transforming the expression

5 A corner has been cut off this rectangle.
Find the area of the piece that remains blue.

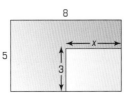

This spread will show you how to:
▶▶ Begin to distinguish the different roles played by letter symbols in equations, formulae and functions.
▶▶ Substitute integers into simple formulae.

KEYWORDS

Function	Formulae
Expression	Substitute
Formula	

In the linear equation $7 + 3x = 1$, x has a particular value that can be found.

The value of x is always the same: $x = {}^-2$.
In the function $y = 7 + 3x$, you can choose a value of x and calculate the value of y.

The value of y changes with different values of x.

▶ An equation can be solved to find a particular value.
▶ A function links two or more unknowns to each other.

You can substitute negative values into functions.

You need to know the rules for multiplying and dividing with two negative numbers.

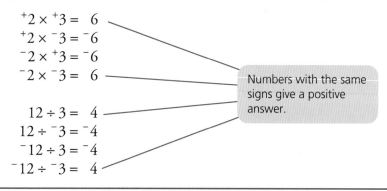

$^+2 \times {}^+3 = 6$
$^+2 \times {}^-3 = {}^-6$
$^-2 \times {}^+3 = {}^-6$
$^-2 \times {}^-3 = 6$

$12 \div 3 = 4$
$12 \div {}^-3 = {}^-4$
$^-12 \div 3 = {}^-4$
$^-12 \div {}^-3 = 4$

Numbers with the same signs give a positive answer.

Which value of x gives the function $y = \dfrac{2x - 4}{1 - 2x}$ the larger value, $x = 1$ or $x = {}^-1$?

It is easier to work with one line at a time.

When $x = 1$: $\quad 2x - 4 = 2 \times 1 - 4 = {}^-2$

$\qquad 1 - 2x = 1 - 2 \times 1 = 1 - 2 = {}^-1$

$y = \dfrac{{}^-2}{{}^-1} = {}^-2 \div {}^-1 = 2$

When $x = {}^-1$: $\quad 2x - 4 = 2 \times {}^-1 - 4 = {}^-2 - 4 = {}^-6$

$\qquad 1 - 2x = 1 - 2 \times {}^-1 = 1 + 2 = 3$

$y = \dfrac{{}^-6}{3} = {}^-6 \div 3 = {}^-2$

The function is larger when $x = 1$.

Exercise A2.5

1 Work out the value of this expression:
$3p^2 - 2p + 1$
when:

  a $p = 1$ **b** $p = {}^-1$
  c $p = 3$ **d** $p = {}^-5$
  e $p = 10$ **f** $p = {}^-12$

2 Work out the value of each expression when $a = 5$, $b = {}^-3$, $c = 2$. Write down the expressions that have a value of 15.

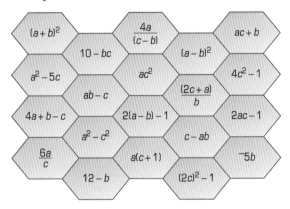

Hexagon cells containing:
$(a + b)^2$, $\dfrac{4a}{(c - b)}$, $ac + b$
$10 - bc$, $(a - b)^2$
$a^2 - 5c$, ac^2, $4c^2 - 1$
$ab - c$, $\dfrac{(2c + a)}{b}$
$4a + b - c$, $2(a - b) - 1$, $2ac - 1$
$a^2 - c^2$, $c - ab$
$\dfrac{6a}{c}$, $a(c + 1)$, ${}^-5b$
$12 - b$, $(2c)^2 - 1$

3 In each function, find the value of y when $x = {}^-3$ and when $x = 3$, and say which gives the larger value.

  a $y = \dfrac{2x + 3}{3}$

  b $y = \dfrac{x - 1}{x + 1}$

  c $y = \dfrac{12 - x}{x}$

  d $y = \dfrac{2x - 2}{1 - x}$

  e $y = \dfrac{2x + 10}{10 - 2x}$

  f $y = 2x - \dfrac{(3x + 1)}{(x - 2)}$

4 Find the value of these expressions when $t = {}^-4$:

  a $4t + 1$ **b** $5 - 4t$
  c $3(t - 1)$ **d** $4 - 3t$
  e $t(8 - t)$ **f** $(t + 3)(t - 1)$

5 In this hexagon there are nine diagonals.

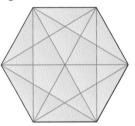

  a How many diagonals can be drawn in a pentagon?

  ▶ The number of diagonals for a polygon with any number of sides is given by the formula:

$$d = \frac{n(n - 3)}{2}$$

    where d = number of diagonals
    and n = number of sides.

  b Use this formula to check your answer to part **a**.

  c Show how this rule explains how many diagonals can be drawn in a triangle ($n = 3$).

  d Use the rule to state how many diagonals there are in:

    i   an octagon
    ii   a nonagon
    iii a decagon.

6 **Challenge**
You are told that a polygon has 434 diagonals when they are all drawn in. How many sides has the polygon?

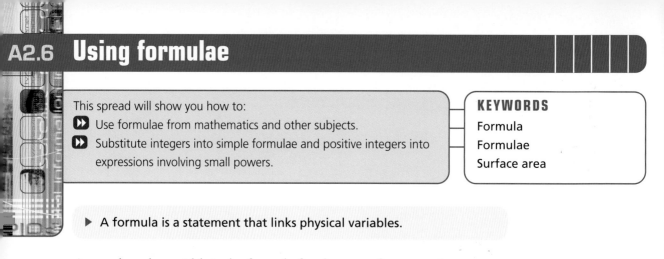

This spread will show you how to:
- ▶▶ Use formulae from mathematics and other subjects.
- ▶▶ Substitute integers into simple formulae and positive integers into expressions involving small powers.

KEYWORDS
Formula
Formulae
Surface area

▶ A formula is a statement that links physical variables.

Area = length × width is the formula for the area of a rectangle.

When you use formulae you must use the correct units of measurement.

In the examples on these pages:

- ▶ Velocity, v, is measured in metres per second, written ms^{-1} or m/s.
- ▶ Acceleration, a, is measured in metres per second per second, written ms^{-2} or m/s^2.
- ▶ Distance, S, is measured in metres.
- ▶ Time, t, is measured in seconds.

example

A large boulder rolls down a chute with a constant acceleration of 2 ms^{-2}.
The distance it travels in a period of time is given by the formula:

$$S = at^2$$

where S is the distance, a is the acceleration and t is the time taken.

a Find the distance travelled after:
  i 3 seconds **ii** 7 seconds **iii** $\frac{1}{2}$ a second

b If the slope is 200 m long, how long will it take the boulder to reach the end?

a You know that $a = 2 \text{ ms}^{-2}$ and that the distance, S, is measured in metres.
  i When $t = 3$, $S = 2 \times 3 \times 3 = 18$ metres
  ii When $t = 7$, $S = 2 \times 7 \times 7 = 98$ metres
  iii When $t = \frac{1}{2}$, $S = 2 \times \frac{1}{2} \times \frac{1}{2} = \frac{1}{2}$ metre.

b You know that $a = 2 \text{ ms}^{-2}$ and that the distance, S, is 200 m.
  $200 = 2t^2$
  $100 = t^2$
  $100 = 10 \times 10 = 10^2$ so $t = 10$ seconds.

Exercise A2.6

1 A boy drops a stone over the edge of a cliff.
The stone takes 3 seconds before it splashes
into the sea at the bottom of the cliff.

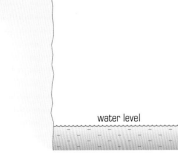

cliff

a Use the formula $v = at$ to find the velocity (speed)
of the stone as it splashes into the sea.
b The distance travelled, S metres, is given by the
formula $S = \frac{1}{2}at^2$.
How high was the cliff above sea level?

water level

> v = velocity (speed) in m/s
> a = acceleration = 10 m/s²
> t = time in seconds = 3 seconds

2 Use the formula $v = at$ to find:

a v if $a = 2$ m/s² and $t = 10$ seconds   **b** v if $a = 4$ m/s² and $t = 8$ seconds
c t if $a = 5$ m/s² and $v = 50$ m/s   **d** t if $a = 2$ m/s² and $v = 45$ m/s.

3 Use the formula $S = \frac{1}{2}at^2$ to find:

a S if $a = 3$ m/s² and $t = 5$ seconds   **b** S if $a = 2$ m/s² and $t = 4$ seconds
c a if $S = 200$ m/s and $t = 10$ seconds   **d** t if $S = 36$ m and $a = 8$ m/s².

4 Find the number midway between:
   a 8 and 12   **b** 15 and 30   **c** 13 and 37   **d** 20 and 84

e The formula $m = \dfrac{a + b}{2}$ gives you the middle number (where m is the

middle number and a and b are the two numbers).
Use this formula to check your answers to parts **a–d** and find the
number midway between 86 and 322.

5 This large carton contains boxes of the breakfast cereal
Fruit Krispies, and is made in four different sizes.

> Volume = LBH
> Surface area = $2LH + 2LB + 2BH$

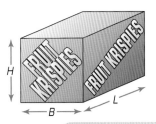

a Explain how the formula for surface area works.
b Complete the table, using the formula provided.

cc means cubic centimetres

Carton	L (cm)	B (cm)	H (cm)	Volume (cc)	Surface area (cm²)
A	60	40	30		
B	80	30	45		
C	75	40	40		
D		25	20	30 000	

You should know how to ...

1 Simplify or transform linear expressions by collecting like terms.

2 Multiply a single term over a bracket.

3 Substitute integers into simple formulae.

4 Represent problems in algebraic form using correct notation.

Check out

1 The number in each cell is made by adding the numbers in the two cells below it.

Fill in the missing expressions.

Write the expressions as simply as possible.

?

$6u - 2t$	$3u - 4t$

?	$3u$	^-4t

2 Simplify these expressions:

a $3(x + 4)$

b $x(y + z)$

c $4(3t - 2)$

d $12 - (x - 5)$

e $3(2x - 1) + 5(x - z)$

3 Find the value of each expression when $x = {}^-3$:

a $4x + 5$

b $3 - 2x$

c $6(x - 1)$

d $x^2 + 1$

e $\dfrac{x}{2x + 3}$

f $\dfrac{x - 1}{x + 1}$

4 This pattern grows by adding squares:

a Copy and complete this table for the first four terms of this sequence

Pattern number	1	2	3	4
Number of squares	1	4		
Number of extra squares	1			

b Use the third pattern to prove that
$1 + 2 + 3 + 2 + 1 = 3^2 = 9$.

c Draw the next pattern in the sequence.

d Use your pattern to prove that
$1 + 2 + 3 + 4 + 3 + 2 + 1 = 4^2 = 16$.

Measures and measurements

This unit will show you how to:

▶▶ Use units of measurement to estimate, calculate and solve problems in everyday contexts involving length, area, volume, capacity and mass.

▶▶ Know rough metric equivalents of imperial measures in daily use.

▶▶ Deduce and use formulae for the area of a triangle, parallelogram and trapezium.

▶▶ Calculate areas of compound shapes made from rectangles and triangles.

▶▶ Know and use the formula for the volume of a cuboid.

▶▶ Calculate volumes and surface areas of cuboids and shapes made from cuboids.

▶▶ Solve more demanding problems and investigate in the contexts of perimeter, area and volume, and measures.

▶▶ Solve more complex problems by breaking them into smaller steps or tasks, choosing and using efficient techniques and resources.

I'm afraid you can only take 32 kg and use 6.2 m^3, madam. Your luggage is far too heavy.

CHECK IN

Most objects have volume and weight.

Before you start

You should know how to ...

1 Choose appropriate metric measures of length.

2 Use the formula for the area of a rectangle.
▶ Area of rectangle = length × width

3 Recognise and describe features of common 3-D shapes.

Check in

1 What units would you use to measure:
  a the height of a tall tree
  b the length of a fly?
  Choose from mm, cm, m and km.

2 Find the area of these rectangles:
  a 5m
     3m
  b 7.3cm
     3.1cm

3 How many faces, edges and vertices does this shape have?

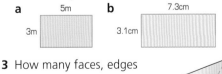

Using metric units

This spread will show you how to:

▶▶ Use units of measurement to estimate, calculate and solve problems in everyday contexts involving length, area, volume, capacity and mass.

KEYWORDS
Square millimetre
Square centimetre
Square metre   Litre
Cubic millimetre
Cubic centimetre
Cubic metre

You measure lengths or distances using mm, cm, m or km.

▶ 10 mm = 1 cm
▶ 100 cm = 1 m
▶ 1000 m = 1 km

Area is the measure of the amount of space a flat shape covers.
Area has two dimensions. You measure it using mm^2, cm^2, m^2 or km^2.

▶ 1 mm^2 is about the size of a pin point.
▶ 1 cm^2 is about the size of your little fingernail.
▶ 1 m^2 is about half the size of a door.

The ^{2} shows there are two dimensions.

Most everyday objects such as a bag or a phone have three dimensions: 3-D.

▶ Volume is the measure of the amount of space a 3-D shape takes up.

You measure volume using:

mm^3

cm^3

m^3

km^3

Size of a ...
... grain of sugar   ... dice   ... washing machine   ... mountain

Capacity is the amount of liquid a 3-D shape can hold.
You measure it using millilitres (ml) or litres (l).

▶ 1000 ml = 1 l.

1 ml is a few drips from a tap.
A medium bottle of drink may hold 1 l.

The smaller the unit the more of them you need to measure a length, so you multiply when you convert to smaller units:
3.3 kg = 3.3 × 1000 g = 3300 g

The mass of an object is the measure of how much it weighs.
You measure mass in grams or kilograms.

▶ 1000 g = 1 kg

A piece of paper weighs about 1 g.
A bag of sugar weighs 1 kg.

Exercise S2.1

1 Convert into metres:
  a 300 cm
  b 46 cm
  c 184 cm

2 Convert into kilograms:
  a 4000 g
  b 3500 g
  c 700 g

3 Convert into litres:
  a 3000 ml
  b 4500 ml
  c 340 ml

4 Suggest appropriate metric units to measure:
  a the height of your school.
  b the distance to the North Pole.
  c the capacity of a swimming pool.
  d the area of a mobile phone display.
  e the mass of an apple.
  f the length of a fly.
  g the capacity of a minibus.
  h the mass of a washing machine.
  i the area of the classroom floor.

5 The Cool-Dude factory makes skateboards. Each deck uses 71.6 cm of wood.
  a How many decks can be made from a 10 m piece of wood?
  b Each skateboard uses 123 cm of tape. In one day the factory makes 4370 skateboards.
    How many kilometres of tape are used in the day?

6 In Tusco a 3 litre bottle of Coke costs £1.59. A 340 ml can costs 39p.
  a Which is the best value and why?
  b Why might you still buy the can?

7 The diagram is a plan of the floor of a school hall.

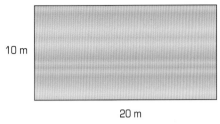

10 m

20 m

  a For an assembly the hall is filled with chairs.

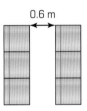

55 cm

55 cm — Each chair needs this amount of space.

You need a gap of 0.6 m between rows of seats.

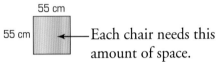

0.6 m

How many chairs can you fit in the school hall?

  b How many chairs can you fit in the school hall if they are arranged in two blocks?
    You need to leave a border of 1 m around each block.

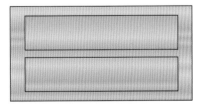

  c Choose your own size of hall. How many chairs can you fit in the hall?

This spread will show you how to:

▶▶ Know rough metric equivalents of imperial measures in daily use.

KEYWORDS
Imperial   Yard
Foot   Equivalent

Most people around the world use metric units.

In Britain many people still use imperial units.

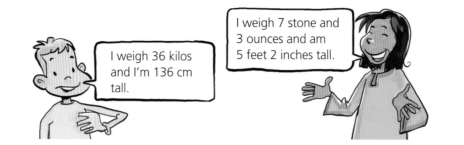

I weigh 36 kilos and I'm 136 cm tall.

I weigh 7 stone and 3 ounces and am 5 feet 2 inches tall.

The imperial units still commonly in use are:

Length	Mass	Capacity
12 inches (in) = 1 foot (ft)	16 ounces (oz) = 1 pound (lb)	8 pints (pt) = 1 gallon
3 feet = 1 yard (yd)	14 pounds = 1 stone (st)	Milk is measured in pints.
1760 yards = 1 mile		

You need to know the metric equivalents of these measures.

Length
▶ 1 inch ≈ 2.5 cm (your ruler may illustrate this!)
   1 foot ≈ 30 cm
   1 yard ≈ 90 cm (about a metre)
▶ 1 mile ≈ 1.6 km (it takes about 20 minutes to walk)
   1 km ≈ $\frac{5}{8}$ mile (this is very useful in practice)

Mass
▶ 1 kg ≈ 2 lb (2.2 lb to be precise – see a bag of sugar!)

Capacity
▶ 1 pint ≈ $\frac{1}{2}$ litre (568 ml to be precise)
   1 litre ≈ 1.75 pints

The imperial system was developed using practical measures such as the length of a foot. The metric system was developed using scientific measures – splitting specific measures into 10s and 100s. So the equivalents are not exact, just rough estimates.

example

A large bottle contains 2 litres of water.
How many pint glasses can you fill from the bottle?

....

1 litre ≈ 1.75 pints, so 2 litres ≈ 2 × 1.75 pints = 3.5 pints.
You can fill 3 pint glasses from the bottle.

Exercise S2.2

1 Copy and complete this table of metric equivalents.

Imperial measurement	Metric unit	Metric equivalent
2 feet	centimetres	60 cm
6 feet		
3 pints		
1 gallon		
25 yards		
100 yards		
10 miles		
174 miles		

2 Copy and complete this table of imperial equivalents.

Metric measurement	Imperial unit	Imperial equivalent
4 kilograms		
3 litres		
150 centimetres		
18 kilometres		
279 km		

3 Measure your height in feet and inches.
Convert your result to metres and centimetres.

4 How many pounds does your school bag weigh?
How many kilograms is this?

5 How long will it take you to run a mile?

6 The supermarket sells strawberries for £2.50 per kilogram. The corner shop sells strawberries for £1 per pound.
Which is the cheapest? How can you tell?

7 1 pint of Coke in a restaurant is £1.30.
A 2 litre bottle costs £1.39.
Which is the best value and why?

8 a The distance to Hathersage is given as 6 miles.
Do you think this is correct to:
the nearest mile, the nearest yard or the nearest quarter mile?
b Change the sign to show distances in kilometres.

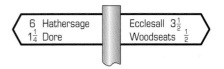
6 Hathersage   Ecclesall $3\frac{1}{2}$
$1\frac{1}{4}$ Dore   Woodseats $\frac{1}{2}$

Area of a triangle and a parallelogram

KEYWORDS

Triangle   Parallelogram
Perpendicular   Deduce
Trapezium

The area of a rectangle is found using the formula:

▶ **Area of rectangle = length × width**

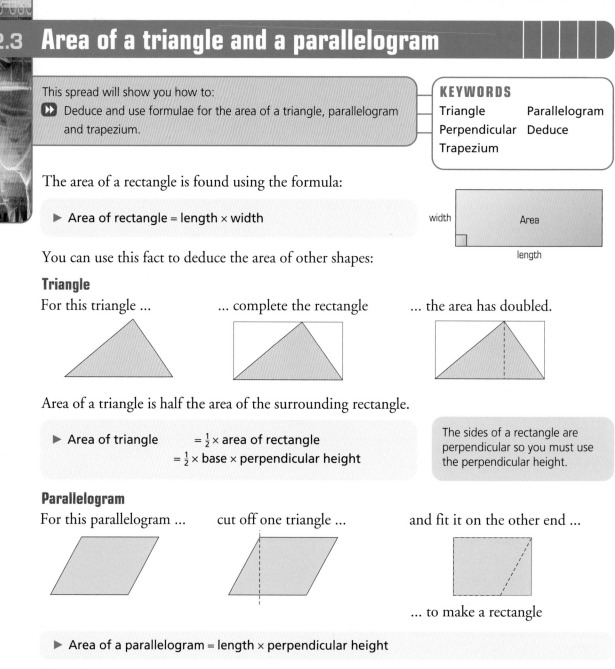

width

Area

length

You can use this fact to deduce the area of other shapes:

Triangle

For this triangle ...   ... complete the rectangle   ... the area has doubled.

Area of a triangle is half the area of the surrounding rectangle.

▶ **Area of triangle** $= \frac{1}{2} \times$ **area of rectangle**
              $= \frac{1}{2} \times$ **base × perpendicular height**

The sides of a rectangle are perpendicular so you must use the perpendicular height.

Parallelogram

For this parallelogram ...   cut off one triangle ...   and fit it on the other end ...

... to make a rectangle

▶ **Area of a parallelogram = length × perpendicular height**

Trapezium

A trapezium can be split up into a rectangle and triangles:

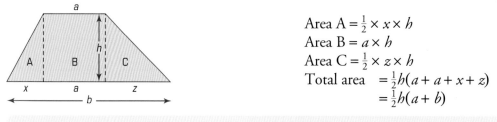

Area A $= \frac{1}{2} \times x \times h$
Area B $= a \times h$
Area C $= \frac{1}{2} \times z \times h$
Total area $= \frac{1}{2}h(a + a + x + z)$
           $= \frac{1}{2}h(a + b)$

▶ **Area of a trapezium = $\frac{1}{2}$ (sum of the parallel sides) × distance between**

Exercises S2.3

1 Calculate the area of these triangles:

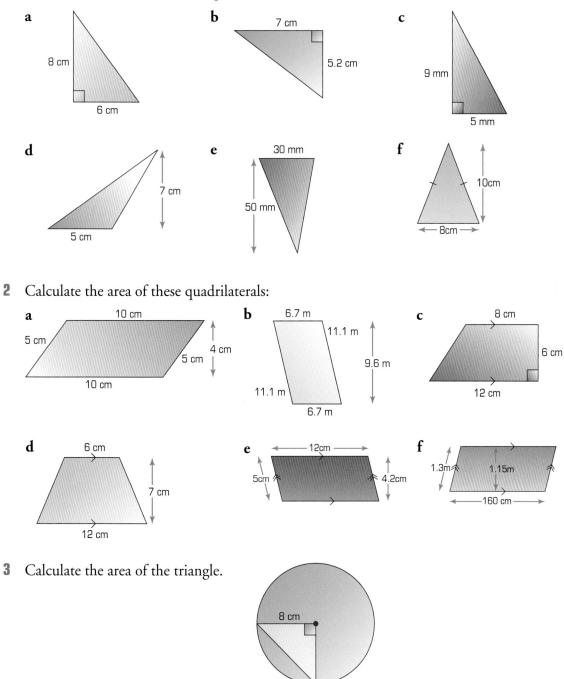

a

8 cm
6 cm

b

7 cm
5.2 cm

c

9 mm
5 mm

d

7 cm
5 cm

e

30 mm
50 mm

f

10cm
8cm

2 Calculate the area of these quadrilaterals:

a

10 cm
5 cm
4 cm
5 cm
10 cm

b

6.7 m
11.1 m
9.6 m
11.1 m
6.7 m

c

8 cm
6 cm
12 cm

d

6 cm
7 cm
12 cm

e

12cm
5cm
4.2cm

f

1.3m
1.15m
160 cm

3 Calculate the area of the triangle.

8 cm

4 **a** Sketch three different right-angled triangles with an area of 12 cm^2.
b Sketch three different non-right-angled triangles with an area of 12 cm^2.

Compound areas

This spread will show you how to:
▶▶ Use formulae for the area of a triangle, parallelogram and trapezium.
▶▶ Calculate areas of compound shapes made from rectangles and triangles.

A trapezium is a made up of rectangles and triangles.

You can find the areas of compound shapes by splitting them into rectangles and triangles.

▶ Area of a rectangle = length × width
▶ Area of a triangle = $\frac{1}{2}$ base × perpendicular height

example

Find the area of this shape.

All length are in cm so all areas are in cm^2.
Split the shape into three parts:

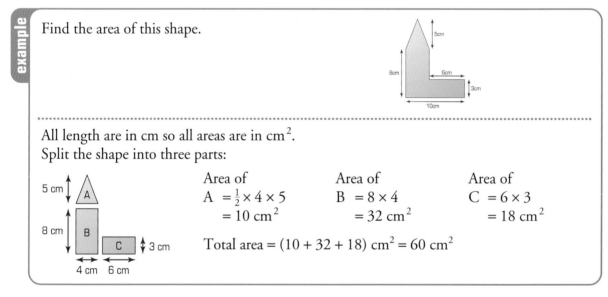

Area of
A $= \frac{1}{2} \times 4 \times 5$
   $= 10 \text{ cm}^2$

Area of
B $= 8 \times 4$
   $= 32 \text{ cm}^2$

Area of
C $= 6 \times 3$
   $= 18 \text{ cm}^2$

Total area $= (10 + 32 + 18) \text{ cm}^2 = 60 \text{ cm}^2$

When you solve problems you must use the correct units:
▶ Lengths are measured in mm, cm, m, km.
▶ Areas are measured in mm^2, cm^2, m^2, km^2.

The ^{2} shows that area has 2 dimensions.

example

The area of this parallelogram is 30 m^2.
Find the length marked x.

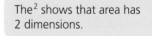

A = 30 m^2 5 m

The area is in m^2 and lengths are in m.
Area = base × perpendicular height
       $30 = x \times 5$
so $x = 6$ m

Exercise S2.4

1 Find the areas of these compound shapes:

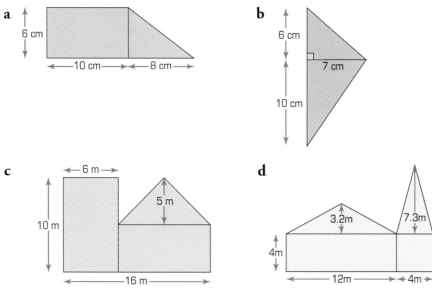

2 This shape is drawn on cm squared paper. Calculate the area of the shaded shape.

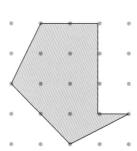

3 **a** This parallelogram has an area 48 m² and base 12 m.

Find *h*.

b This parallelogram has area 20 cm².

Write down the possible values of *b* and *h*.

4 Use cm squared paper.
  a Draw three different triangles with an area of 8 cm².
  b Draw three different parallelograms with an area of 12 cm².
  c Draw three different trapezia with an area of 12 cm².

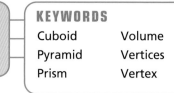

This spread will show you how to:

▶▶ Know and use the formula for the volume of a cuboid.

▶▶ Calculate volumes of cuboids and shapes made from cuboids.

KEYWORDS

Cuboid	Volume
Pyramid	Vertices
Prism	Vertex

Most everyday objects have three dimensions (3-D):
length, width and height.
These are common 3-D shapes:

Cube	Cuboid	Prism	Pyramid
All faces are squares.	All faces are rectangles.	The cross-section is constant.	The base tapers to a point.

vertex

▶ The volume of a 3-D object is the amount of space it takes up.

▶ You measure the volume in mm³, cm³, m³ or km³.

The 3 shows there are 3 dimensions.

You can find the volume of a cuboid by counting cubes:

There are 4 × 5 = 20 cubes on the first layer.
There are 3 layers.
There are 3 × 20 = 60 cubes altogether.

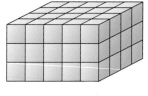

Notice that you multiply the three dimensions together:
3 × 4 × 5 = 60

It is quicker and easier to use the formula:

▶ Volume of a cuboid = length × width × height

example

Find the volume of this cuboid:

6.2cm

8.9 cm

4.5cm

All lengths are in cm so the volume is in cm³.

Volume = length × width × height
= 8.9 × 6.2 × 4.5
= 248.31 cm³

Volume = 248 cm³ to the nearest cm³.

Exercise S2.5

1 Find the volume of these cuboids:

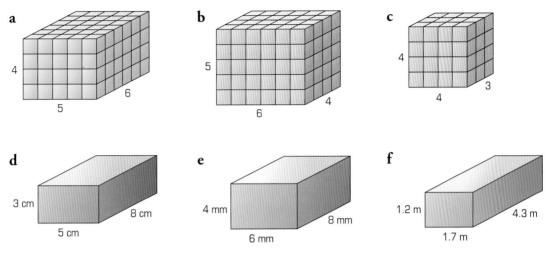

a — 4, 5, 6

b — 5, 6, 4

c — 4, 4, 3

d — 3 cm, 8 cm, 5 cm

e — 4 mm, 8 mm, 6 mm

f — 1.2 m, 4.3 m, 1.7 m

2 Find the volume of these T-shaped girders by splitting them into two cuboids.

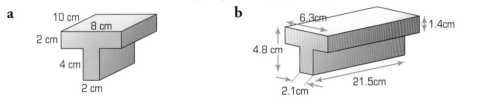

a — 10 cm, 8 cm, 2 cm, 4 cm, 2 cm

b — 6.3cm, 1.4cm, 4.8 cm, 21.5cm, 2.1cm

3 Find the volume in cm³ of this L-shaped girder by splitting it into cuboids.

10 cm

10 cm

32 cm

90 cm

25 cm

4 Boxes of Choco biscuits measure 12.5 cm by 24 cm by 26 cm.

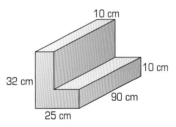

12.5 cm

Choco *Choco*

26 cm

24 cm

The biscuits must stay this way up!

How many boxes of biscuits will fit in this crate:

1 m   **UP ↑**

3 m

2 m

Surface area

This spread will show you how to:
▶▶ Calculate surface areas of cuboids.

KEYWORDS
Cuboid   Net
Surface area

Many boxes start out as flat packs called nets:

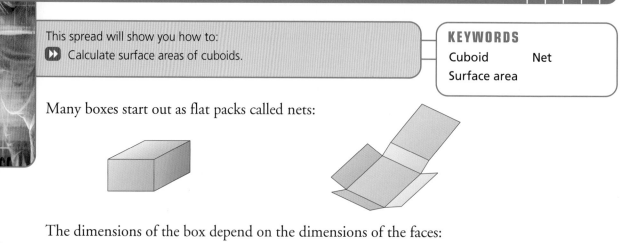

The dimensions of the box depend on the dimensions of the faces:

This net ...

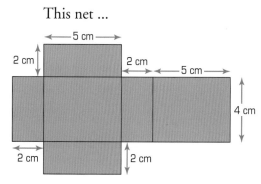

will fold to make this cuboid

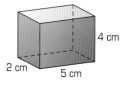

The area of the net gives you the surface area of the cuboid.

▶ **The surface area of a cuboid is the total area of its faces.**

Notice that the faces are in pairs so you only need to work out three areas.

example

Find the surface area of a cuboid measuring 4 cm by 3.5 cm by 3 cm.

All lengths are in cm so areas will be in cm^2.
Sketch the cuboid:

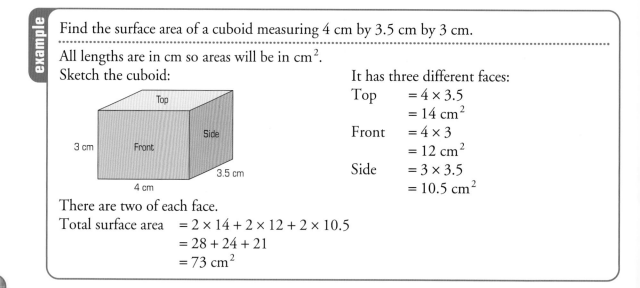

It has three different faces:
Top $= 4 \times 3.5$
         $= 14$ cm^2
Front $= 4 \times 3$
         $= 12$ cm^2
Side $= 3 \times 3.5$
         $= 10.5$ cm^2

There are two of each face.
Total surface area $= 2 \times 14 + 2 \times 12 + 2 \times 10.5$
                     $= 28 + 24 + 21$
                     $= 73$ cm^2

Exercise S2.6

1 Which of these nets will make a cube?

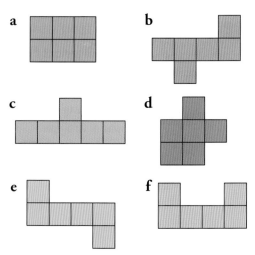

2 Use five multilink cubes to make five different solids.
Sketch your solids.

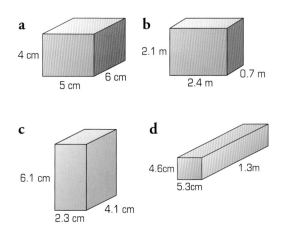

What is the surface area of each of your solids?

3 Work out the surface area of each of these cuboids.

a
4 cm
5 cm
6 cm

b
2.1 m
2.4 m
0.7 m

c
6.1 cm
2.3 cm
4.1 cm

d
4.6cm
5.3cm
1.3m

4 Sketch the net of a cuboid with dimensions 4 cm by 6 cm by 8 cm.
a Calculate the surface area of the cuboid.
b Calculate the volume of the cuboid.

5 Six squares joined together are called a hexomino.

a is a hexomino.

Does it make a cube?

b 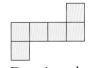 is a hexomino.

Does it make a cube?
c Draw the twelve different hexominoes that will make cubes.

6 Five squares joined together are called a pentomino.
Draw pentominoes which will make an open box.

7 The average size book sold by the online store Book4You.com is
24.5 cm × 20 cm × 1.5 cm.

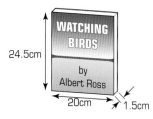

WATCHING BIRDS
by
Albert Ross
24.5cm
20cm
1.5cm

Design a net of a packing box to fit:
a one book
b three books
c ten books.
d Draw an alternative design for ten books.

A3.1 Algebraic functions

This spread will show you how to:

▶▶ Express simple functions in symbols.

▶▶ Represent mappings expressed algebraically.

KEYWORDS

Input | Output
Function | Function machine
Mapping | Linear function

You can write the rule for a function machine using algebra.
The input is the x value and the output is the y value.

In this machine:

The function is $y = x + 6$

The inputs are missing from this machine.

6
11
8
15

The function is $y = \dfrac{x}{2} + 5$

You can work backwards using the inverse functions to find the inputs.

2
12
6
20

6
11
8
15

The inverse function is $x = 2(y - 5)$
The missing inputs are 2, 12, 6 and 20.

When you need to find the functions it can help to use a mapping diagram.

example

For this function machine:

a find the functions

b write the rule for the machine using algebra.

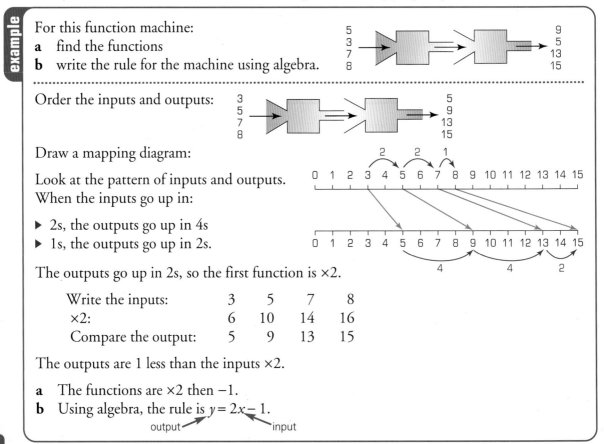

Order the inputs and outputs:

Draw a mapping diagram:

Look at the pattern of inputs and outputs.
When the inputs go up in:

▶ 2s, the outputs go up in 4s

▶ 1s, the outputs go up in 2s.

The outputs go up in 2s, so the first function is ×2.

Write the inputs:	3	5	7	8
×2:	6	10	14	16
Compare the output:	5	9	13	15

The outputs are 1 less than the inputs ×2.

a The functions are ×2 then −1.

b Using algebra, the rule is $y = 2x - 1$.

output ⟶   ⟵ input

Exercise A3.1

1 Find the missing outputs or inputs for each of these function machines:

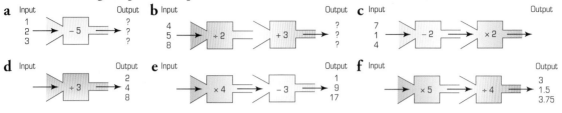

a Input 1, 2, 3 → − 5 → Output ?, ?, ?

b Input 4, 5, 8 → ÷ 2 → + 3 → Output ?, ?, ?

c Input 7, 1, 4 → − 2 → × 2 → Output

d Input → + 3 → Output 2, 4, 8

e Input → × 4 → − 3 → Output 1, 9, 17

f Input → × 5 → ÷ 4 → Output 3, 1.5, 3.75

2 Complete each function machine and match it with one of these functions.

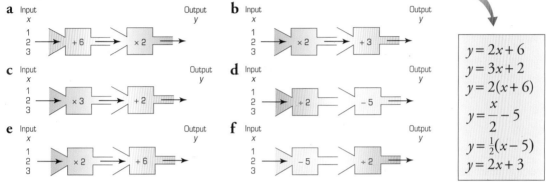

a Input x: 1, 2, 3 → + 6 → × 2 → Output y

b Input x: 1, 2, 3 → × 2 → + 3 → Output y

c Input x: 1, 2, 3 → × 3 → + 2 → Output y

d Input x: 1, 2, 3 → + 2 → − 5 → Output y

e Input x: 1, 2, 3 → × 2 → + 6 → Output y

f Input x: 1, 2, 3 → − 5 → + 2 → Output y

$$y = 2x + 6$$
$$y = 3x + 2$$
$$y = 2(x + 6)$$
$$y = \frac{x}{2} - 5$$
$$y = \tfrac{1}{2}(x - 5)$$
$$y = 2x + 3$$

3 Draw a function machine for each of these algebraic functions:

a $y = 3x - 5$ **b** $y = 2(x - 4)$

c $y = \tfrac{1}{2}x + 5$ **d** $y = \dfrac{x + 4}{3}$

Input   x → [ ] → [ ] → y   Output

4 For each of these function machines, find the rule and write it using algebra.

a Input 1, 2, 3, 4 → [ ] → [ ] → Output 6, 11, 16, 21

b Input 3, 1, 7, 4 → [ ] → [ ] → Output 9, 5, 17, 11

c Input 6, 3, 1, 5 → [ ] → [ ] → Output 10, 4, 0, 8

d Input 5, 11, 3, 8 → [ ] → [ ] → Output 8, 26, 2, 17

5 Challenge

Use two of the functions to make different function machines.

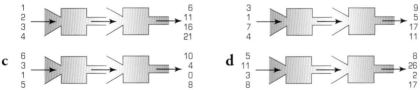

→ + 3 →   → × 5 →   → + 2 →   → − 4 →

a How many different function machines can you make?

b Write each as a linear function.

c Which of them can you convert to a single function machine?

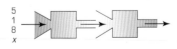

5, 1, 8, x → [ ] → [ ] →

This spread will show you how to:

▶▶ Generate points in all four quadrants and plot the graphs of linear functions, where y is given explicitly in terms of x.

KEYWORDS

Coordinate pairs
Function　　　Graph
Function machine

The input and output values of a function machine form pairs of values.

You can plot the pairs on a coordinate grid.

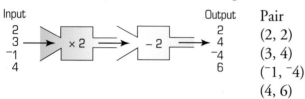

Pair
(2, 2)
(3, 4)
(⁻1, ⁻4)
(4, 6)

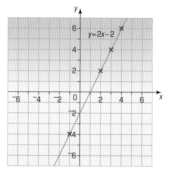

The sequence of points lies on a straight line.

The algebraic function for this machine is $y = 2x - 2$.

This function describes all the values of x and y that lie on the line.

You join the points together to draw the graph of $y = 2x - 2$.

▶ **You can use a function machine to help generate points on the graph of an algebraic function.**

example

Plot the graph of $y = 2x + 3$.

The function $y = 2x + 3$ corresponds to the function machine:

Use values of x to complete a table of values:

x	⁻2	⁻1	0	1	2
y	⁻1	1	3	5	7

Use the coordinate pairs to plot the graph.

Join the points together to form a straight line:

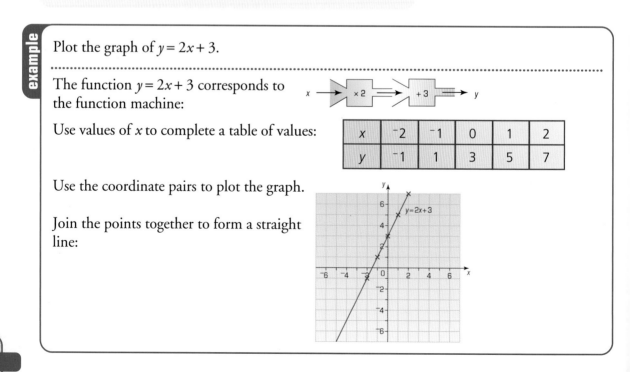

Exercise A3.2

1 For this function:

  a Write down the coordinate pairs made.

  b Write down the linear function for y in terms of x.

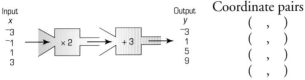

Coordinate pairs

(,)
(,)
(,)
(,)

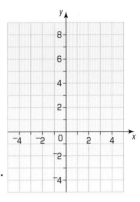

  c Copy the coordinate grid and plot these points.
    Join the points with a straight line to represent the linear function.

2 For this function:

  a Write down the coordinate pairs made.

  b Write down the linear function for y in terms of x.

  c Copy and extend the coordinate grid and plot these points.
    Join the points with a straight line to represent the linear function.

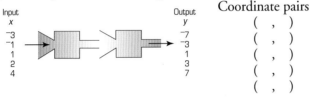

Coordinate pairs

(,)
(,)
(,)
(,)
(,)

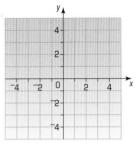

3 For each of these functions:

  ▶ complete the tables of values,

  ▶ plot the coordinate pairs and

  ▶ join up with a straight line to draw the line of the equation.

  a $y = 3x - 1$

x	-2	-1	0	1	2	3
$y = 3x - 1$	-7		-1			

  b $y = 2x - 5$

x	-2	-1	0	1	2	3
$y = 2x - 5$		-7			-1	

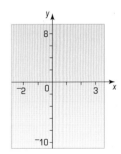

4 This is the line for the function $y = 2x + 4$:

  a At what point does the line cross the y-axis?

  b Copy this grid and line $y = 2x + 4$.

  c On your grid, draw in the lines

    $y = 2x + 1$

    $y = 2x - 2$ (You may want to draw out tables of values.)

  d Look at all of the lines drawn. What can you say about them?
    Compare the equations. What is similar?

  e Write down the intercept for each line (where it crosses
    the y-axis).
    Compare these with each equation. Comment on your findings.

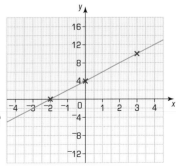

This spread will show you how to:

▶▶ Recognise that equations of the form $y = mx + c$ correspond to straight-line graphs.

KEYWORDS

Gradient   Intercept

Steepness   Function

The graphs of these functions make a pattern.

They all pass through (0, 2) on the *y*-axis.

The graphs get steeper as the multiplier gets bigger.

Look at the equations:

$$y = \boxed{\tfrac{1}{2}x} \quad \boxed{+ 2}$$
$$y = \boxed{x} \quad \boxed{+ 2}$$
$$y = \boxed{2x} \quad \boxed{+ 2}$$
$$y = \boxed{3x} \quad \boxed{+ 2}$$

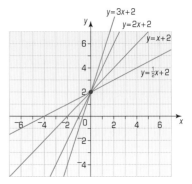

The intercept is (0, 2) for all points.

All the graphs are of the form:

$$y = \ mx \ + c$$

m is a measure of the steepness or gradient   *c* is the *y*-intercept.

▶ A graph with an equation of the form $y = mx + c$:
  ▶ is a straight-line graph
  ▶ crosses the *y*-axis at (0, *c*)
  ▶ gets steeper as *m* gets bigger.

$y = mx + c$ is a linear equation.
(0, *c*) is the *y*-intercept.
m is the gradient of the line.

You only need to know two points to sketch a straight line.

example

Sketch the graph of $y = 2x - 2$.

Choose two *x* values: $x = 0$ and $x = 3$.
Find the corresponding *y* values:
When $x = 0 \quad y = {}^-2$
When $x = 3 \quad y = 6 - 2 = 4$

Two points on the line are $(0, {}^-2)$ and $(3, 4)$.

Join them to sketch the graph.

Check using a third point: $x = 1$:

When $x = 1 \quad y = 2 - 2 = 0$

$(1, 0)$ is on the graph so the graph is correct.

Exercise A3.3

1 a

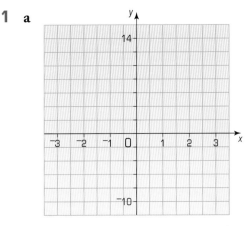

Copy the coordinate grid and by working out the values for y when $x = {}^-3$ and $x = 0$, sketch the graphs of:

   i $y = 3x$
   ii $y = 3x + 4$
   iii $y = 3x - 1$
   iv $y = 3x + 1$

  b Check your lines are correct with the point where $x = 2$.
      Explain what is similar about these graphs.

2 a On a new grid, sketch the graphs of:
   i $y = 2x + 5$
   ii $y = 2x - 5$
   iii $y = 2x + 1$
   iv $y = 2x - 1$

  b Express what is similar about the graphs.

3 a On a new grid, sketch the graphs of:
   i $y = 2x + 3$
   ii $y = x + 3$
   iii $y = {}^-x + 3$
   iv $y = 3x + 3$

  b Explain what is similar about these graphs.

4 On a new grid, sketch and label each of these graphs:
  a $y = 3x + 1$
  b $y = 5x - 2$
  c $y = 2x + 7$
  d $y = 7 - 2x$
  e $y = 3 - 4x$
  f $y = {}^-2 - x$

5 Copy the grids and sketch in the extra graphs. What shape have you made between all of your lines?

  a

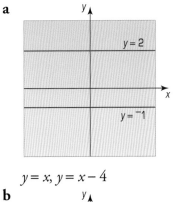

    $y = x, \; y = x - 4$

  b

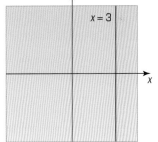

    $y = 6, \; y = 6 - 2x$

  c

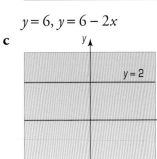

    $y = x + 2, \; y = 4 - x, \; y = {}^-3$

This spread will show you how to:

▶▶ Construct linear functions arising from real-life problems and plot their corresponding graphs.

KEYWORDS

Axis   Axes

Graph   Equation

In Japan £100 is worth 18 000 yen.

It is fairly easy to work out how much 9000 yen is worth: £50.

It is harder to work out how much 5000 yen, or £65 is worth.

A conversion graph will help you to find approximate values.

▶ Draw axes from £0 to £150 and 0 to 20 000 yen.

▶ You know that:
  ▶ £0 is worth 0 yen
  ▶ £100 is worth 18 000 yen

▶ Plot the points:
  (0, 0) and (18 000, 100)

▶ Join them to make a straight line.

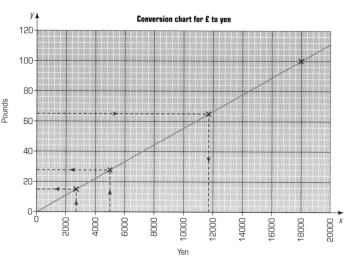

You can use the graph to find exchange values:

▶ Draw a line from the known value to the straight line.
▶ Read off the corresponding value from the other axis.

2700 yen ≈ £15.00

5000 yen ≈ £28.00

£65 ≈ 12 000 yen

The two quantities are linked by the equation:

$$y = \frac{100x}{18\ 000} \qquad \text{where} \qquad y = £ \qquad \text{and} \qquad x = \text{yen.}$$

This simplifies to $\qquad y = \dfrac{x}{180}$

which is of the form $\quad y = mx + c \quad$ (c is 0).

The conversion graph is a straight line with equation $y = \dfrac{1}{180}x.$

Exercise A3.4

1 This graph shows the rate of exchange for £(sterling) into euros in January 2002 when the euro became currency in most EU countries.

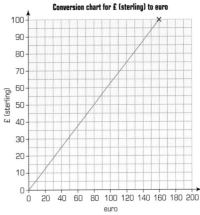

Conversion chart for £ (sterling) to euro

a Copy this chart carefully onto squared paper.
b The × marks the point to show that £100 = 160 euros. Draw the horizontal and vertical lines to show this.
c Use the conversion chart to find the value in euros of:
  i £20   **ii** £80   **iii** £55   **iv** £15
d Use the graph to find the value in £(sterling) of:
  i 40 euros   **ii** 100 euros
  iii 70 euros   **iv** 140 euros

2 This graph is a conversion chart for degrees Fahrenheit to degrees Celsius.

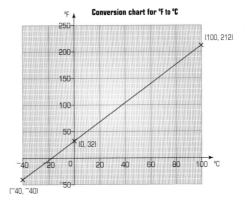

Conversion chart for °F to °C

a Copy this chart carefully onto squared paper.
b The boiling point of water is 100 °C, the freezing point 0 °C. What are these temperatures in °F?
c Change these temperatures into °F:
  i 50 °C   **ii** 25 °C   **iii** 8 °C   **iv** ⁻12 °C
d Change these temperatures into °C:
  i 70 °F   **ii** 50 °F   **iii** 180 °F
  iv 0 °F   **v** ⁻25 °F
e Normal body temperature is 98.4 °F, what is this in °C (approximately)?

3 In continental Europe distances and speed limits are given in km and km per hour (kph).
a Plot a conversion graph for km to miles.
  Use squared paper and sensible scales.
  Use these facts: 0 km = 0 miles
                 160 km = 100 miles
  Plot the points and join up with a straight line.
b Convert these speeds to miles per hour (mph):
  i 20 kph   **ii** 30 kph   **iii** 50 kph   **iv** 70 kph   **v** 130 kph
c Convert these speeds to km per hour (kph):
  i 30 mph   **ii** 40 mph   **iii** 50 mph   **iv** 60 mph   **v** 70 mph

4 Draw a conversion graph to change gallons to litres.
Use the fact that 2 gallons = 9 litres.
Use your graph to estimate the number of gallons in 100 litres and the number of litres in 100 gallons.

This spread will show you how to:
▶▶ Construct linear functions arising from real-life problems and plot their corresponding graphs.
▶▶ Discuss and interpret graphs arising from real-life situations.

KEYWORDS
Construct
Distance–time graph
Gradient

A distance–time graph illustrates a journey by showing how far you have travelled during a period of time:

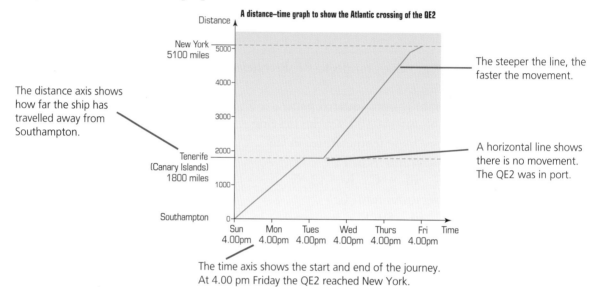

A distance–time graph to show the Atlantic crossing of the QE2

The distance axis shows how far the ship has travelled away from Southampton.

The steeper the line, the faster the movement.

A horizontal line shows there is no movement. The QE2 was in port.

The time axis shows the start and end of the journey.
At 4.00 pm Friday the QE2 reached New York.

example

This distance–time graph shows the journey of a group of friends from Manchester to the Glastonbury festival.

They drove slowly for the first hour until they reached the motorway.

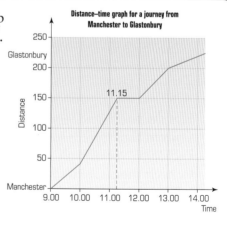

Distance–time graph for a journey from Manchester to Glastonbury

Use the graph to answer these questions:

a What time did they reach the motorway?
b For how long did they stop at the petrol station?
c What time did they arrive at Glastonbury?
d What do you think was happening during the last part of the journey?

a They reached the motorway at 10 am as they started to travel faster.
b They stopped at 11.15 for 45 minutes.
c They arrived at 14.15 (2.15 pm)
d They travelled slowly – there may have been a lot of traffic.

Exercise A3.5

1 This is a distance–time graph for Concorde at its cruising speed.

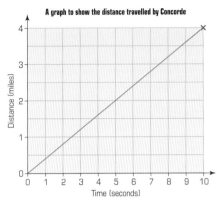

A graph to show the distance travelled by Concorde

a Approximately how far does it travel in 10 seconds?

b How long does it take to travel 1 mile?

c In one minute, how far would Concorde travel?

2 In training, triathletes set targets to swim, cycle and run at constant speeds.
This graph shows a training schedule for 1000 m swimming, cycling and running.

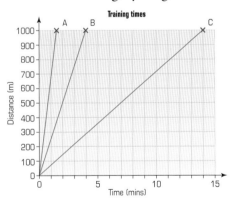

Training times

a Which of these training events is slowest?

b Which of the graphs are for:
  i swimming
  ii cycling
  iii running?

c In training, how long does it take to complete 1000 m in each event?

3 Here are the distance–time graphs for three long distance runners in training.

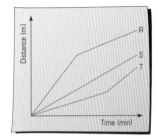

Which runner:

a went slowly at first but then speeded up?

b went at a constant speed?

c travelled the least distance?

4 The London to Edinburgh train leaves London at 11.30 pm.
It arrives in Edinburgh at 7.00 am.
The graph shows the journey.

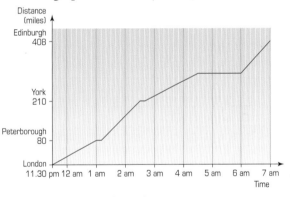

a How long does the train stop in Peterborough?

b When does the train arrive in York?

c What happens between 4.30 and 6.00 am?

d Which part of the journey was the fastest? Which was the slowest?

5 Construct a distance–time graph for this triathlete:
swimming: 1500 m, 25 minutes
cycling: 40 000 m, 40 minutes
running: 10 000 m, 35 minutes

This spread will show you how to: ▶▶ Discuss and interpret graphs arising from real-life situations.	**KEYWORDS** Interpret Deduce Graph

Each day one July the warmest temperature in Cardiff is recorded.
The points are plotted and joined to show the overall picture:

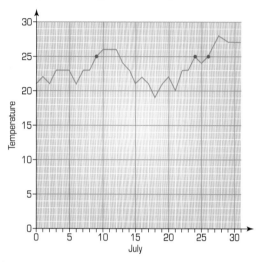

The warmest period was at the end of the month.
The hottest day was 27th July at 28 °C.

The coolest period was between the 14th and
the 22nd July. The coldest day was the 18th July.

Three days had a top temperature of 25 °C:
9th, 24th and 26th July.

You need to be able to interpret graphs showing real-life situations.

example

The graph shows the weight of Joe's lolly left on the
stick as he eats it.
Use the graph to answer these questions:

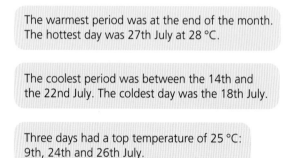

a How long did it take Joe to eat the lolly?
b When did a large piece of lolly fall to the ground?
c Joe took two large bites whilst eating the lolly.
How big were his bites?

....

a After 10 minutes there was no lolly left, so it took Joe
10 minutes to eat.
b A sharp drop in the graph occurred after 7 minutes,
so the piece fell then.
c Two equal-sized drops in weight represent these two
large bites.
They were 20 g each.

Exercise A3.6

1 Sarah drives from Leeds to Bristol on a Wednesday morning, leaving at 8.30 am and arriving in Bristol at 12.30 pm.
The graph shows how much petrol was in the car during the journey.

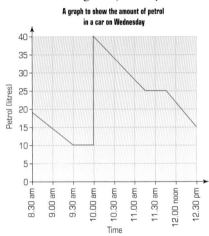

A graph to show the amount of petrol in a car on Wednesday

a Work out how much petrol was in the car when she arrived in Bristol.
b Explain what happened at 10.00 am.
c At which times did Sarah stop and have a break from driving?
d How much petrol did she use in her journey from Leeds to Bristol?

2 To cook frozen peas, you place them in boiling water and then bring the water back to the boil.

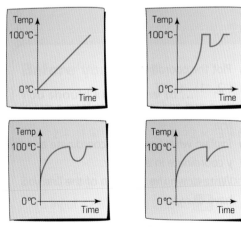

Which of these diagrams best describes frozen peas being cooked? Explain your answer in words.

3 This graph shows the temperature for a sunny day in March in London.

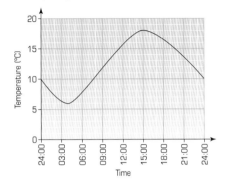

a What time was it warmest?
b What was the coldest temperature?
c For how many hours did the temperature rise?
d When was the temperature 10 °C?

4 Hot water was placed in three test tubes and the temperature recorded over two hours:
▶ X had no wrapping
▶ Y was placed in a freezer
▶ Z was wrapped in plastic foam.
The graph shows the temperature pattern for test tube X.

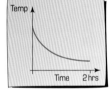

Sketch the graphs for test tubes Y and Z.
How they are different?

5 This graph shows the number of people at school during a normal school day.

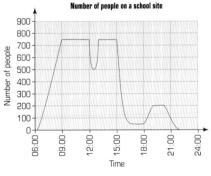

Number of people on a school site

Explain what the graph shows.

This spread will show you how to:
▶▶ Read and write positive integer powers of 10.
▶▶ Multiply and divide numbers by 0.1 and 0.01.
▶▶ Use the power key on a calculator.

KEYWORDS
Power
Power key
Billion

The decimal system is based upon powers of 10:

1 ten	= 10	= 10	= 10^1
1 hundred	= 100	= 10×10	= 10^2
1 thousand	= 1000	= $10 \times 10 \times 10$	= 10^3
10 thousand	= 10 000	= $10 \times 10 \times 10 \times 10$	= 10^4
100 thousand	= 100 000	= $10 \times 10 \times 10 \times 10 \times 10$	= 10^5
1 million	= 1 000 000	= $10 \times 10 \times 10 \times 10 \times 10 \times 10$	= 10^6
1 billion	= 1 000 000 000		= 10^9

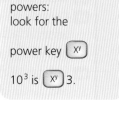

You can use a calculator to input powers:
look for the
power key $\boxed{x^y}$
10^3 is $\boxed{x^y}$ 3.

The table shows the value of each of the digits in the number 3 426 126.

	$\times 10$	$\times 10$	$\times 10$	$\times 10$	$\times 10$	$\times 10$	
10^6 Millions	10^5 Hundred thousands	10^4 Ten thousands	10^3 Thousands	10^2 Hundreds	10^1 Tens	Units
3	4	2	6	1	2	6

← Each place is 10 times bigger

÷ 10   ÷ 10   ÷ 10   ÷ 10   ÷ 10   ÷ 10

The digit 6 stands for
6 thousands = $6 \times 10^3 = 6 \times 1000 = 6000$

→ Each place is 10 times smaller

You can use place value to multiply and divide by 0.1 and 0.01.

You know that:

▶ Multiplying by 0.1   = Multiplying by $\frac{1}{10}$ = Dividing by 10
7×0.1 = $7 \times \frac{1}{10}$ = $7 \div 10$ = 0.7

▶ Multiplying by 0.01   = Multiplying by $\frac{1}{100}$ = Dividing by 100
7×0.01 = $7 \times \frac{1}{100}$ = $7 \div 100$ = 0.07

Multiplying is the inverse of dividing:
if $7 \times 0.1 = 0.7$ then $0.7 \div 0.1 = 7$

×0.1
7   0.7
÷ 0.1

It follows that:

▶ Dividing by 0.1   = Dividing by $\frac{1}{10}$ = Multiplying by 10
$7 \div 0.1$ = $7 \div \frac{1}{10}$ = 7×10 = 70

▶ Dividing by 0.01   = Dividing by $\frac{1}{100}$ = Multiplying by 100
$7 \div 0.01$ = $7 \div \frac{1}{100}$ = 7×100 = 700

Ask: how many tenths are there in 7?

Exercise N3.1

1 Calculate:
   a $23 \div 10$ **b** 4.56×100
   c $394 \div 100$ **d** $0.4 \div 10$
   e 3.07×10 **f** 0.034×100
   g $0.27 \div 1000$

2 Samantha and Duane are investigating how to multiply numbers by 0.1.
   Samantha thinks it's easier to work the sums out by dividing by a different number in her head.
   Duane thinks it's easier to use a calculator.
   a What number is Samantha dividing by?
   b Investigate how to:
      ▸ multiply by 0.01
      ▸ divide by 0.1
      ▸ divide by 0.01
   using Samantha's method for mental calculation.

3 Calculate:
   a 5×0.1 **b** 0.01×4
   c 10×1.2 **d** $8 \div 0.1$
   e $9 \div 0.01$ **f** 67×0.01
   g $^-25 \div 0.01$ **h** 0.7×0.01

4 **Puzzle**
   Work out these expressions and put the answers in order from highest to lowest.
   Do you agree with what you spell?

   S $34 \div 1000$ N 490×0.01
   M $651 \div 0.01$ I 5×100
   T 72×10 G 6×0.1
   Y $^-63 \div 0.1$ I 12×0.01
   S $^-9 \times 10$ I $112 \div 10$
   A $^-350 \div 100$ L $3 \div 0.01$
   L 10×93 U 5.4×1000
   E 0.3×0.1 P 35×10
   Y $10\,640 \div 100$

5 Copy and complete:
   a $0.7 \times ? = 70$
   b $0.3 \times ? = 0.03$
   c $? \div 0.1 = 64$
   d $7.8 \times ? = 0.078$
   e $^-4.5 \div ? = ^-450$
   f $^-0.23 \div ? = 2.3$

6 **Puzzle**
   In a multiplication trail you choose a number from each row which multiplies to make the target number.

4	6	8
9	8	2
0.1	0.01	10

0.32

   For example: $4 \times 8 = 32$
                 $32 \times 0.01 = 0.32$

   Identify which numbers multiply to make the target number in these multiplication trails.

   a

6	90	$^-12$
4	$^-6$	$^-0.3$
0.1	10	0.01

0.36

   b

$^-16$	96	24
$^-0.3$	120	18
$^-0.01$	$^-0.1$	10

$^-28.8$

7 Put the answers to each of these calculations in order from lowest to highest:
   4.6×10^2 $5.12 \div 0.01$
   $4520 \div 10$ 0.052×10^4
   $4.07 \div 10^2$ 0.428×10^3

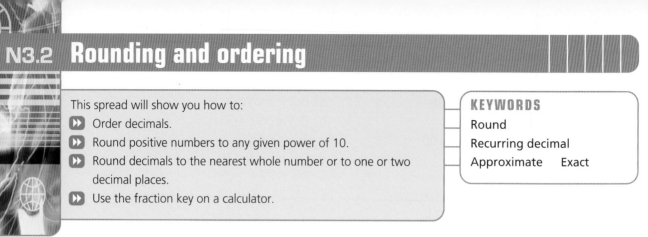

This spread will show you how to:
- ►► Order decimals.
- ►► Round positive numbers to any given power of 10.
- ►► Round decimals to the nearest whole number or to one or two decimal places.
- ►► Use the fraction key on a calculator.

KEYWORDS
Round
Recurring decimal
Approximate   Exact

Sometimes it is more appropriate to give the approximate size of a number or quantity than its exact size.

The crowd was about 120 000.

Attendance 121 367

This car travels about 42 miles per gallon.

Supersupi Sports GXLi

Max speed	**123 mph**
MPG	**41.65**
BHP	**1203**

You can round decimals to an appropriate decimal place.

example

Round 5.3781 to: (a) 1 decimal place   (b) 2 decimal places

a   To round to the nearest tenth (1 dp) look at the hundredths digit:
5.3781 is between 5.3 and 5.4
5.3781 ≈ 5.4 (to 1 dp)

5.3 ├──────────────┤ 5.4
                    ↑
                 5.3781

b   To round to the nearest hundredth (2 dp) look at the thousandths digit:
5.3781 is between 5.37 and 5.38
5.3781 ≈ 5.38 (to 2 dp)

5.37 ├──────────────┤ 5.38
                    ↑
                 5.3781

You can order decimals by considering the value of each decimal place in turn.

example

Order 0.625, 0.66, 0.ȯ6 and 0.6

The first dp are all the same.
Comparing the second dp: 0.6, 0.625, 0.ȯ6 and 0.66
Comparing the third dp: 0.6, 0.625, 0.66 and 0.ȯ6

In 0.ȯ6 the dot shows that the digit 6 recurs.
0.ȯ6 = 0.666 ...

├─────┼─────┼──────┼┼─┤
      ↑   ↑   ↑ ↑
    0.6   0.625   0.66 0.ȯ6

Exercise N3.2

1 Round each of these numbers to the nearest:

 a whole number

 b one decimal place

 c 2 decimal places

  i 13.284 **ii** 1.654

  iii 173.799 **iv** 3.0734

2 Here is the manufacturer's data on a high performance car.
The items in red are features that you would like to be low – the items in black you would like to be high.

Performance Data:	
Engine size	2174 cc
Top speed	152 mph
BHP	174
0–60 mph	8.4 secs
MPG	37.2
Price	£14 995

 a Round the numbers to an appropriate degree of accuracy to enhance the car's selling points.
 For example:
 engine size = 2200 (nearest hundred) ✗
   = 2000 (nearest thousand) ✓

 b Write a short advert using your numbers to promote the car.
 For example: 'The engine is only a 2000 cc, so the insurance will be low ... '

3 **Puzzle**
Place these numbers in order from lowest to highest to spell a team going backwards.

A 0.01857	N −0.036	E 0.02
T ⁻0.037	D ⁻0.041	O 0.29
R 0.3	M 0.017	E ⁻0.0375
H 0.17	T 0.172	R 0.019
H 0.0186	U ⁻0.03	I ⁻0.0364

4 Find the number that lies exactly halfway between each of these pairs of numbers:

 a 2.4 and 3.7

 b 0.35 and 0.69

 c ⁻2.6 and ⁻1.4

 d ⁻3.84 and 3.05

5 **Investigation**
Use the fraction key $\boxed{a\frac{b}{c}}$ on your calculator to convert these fractions into their decimal equivalents:

 a $\frac{2}{3}$ **b** $\frac{5}{16}$ **c** $\frac{7}{9}$ **d** $\frac{7}{12}$ **e** $\frac{23}{40}$ **f** $\frac{7}{11}$

  i Write the decimals to 2 decimal places. Write any recurring decimals using the correct notation

  ii Use your calculator to find the fraction that is the same as 0.272 727 ...

  iii Investigate other fractions that give recurring decimals such as 0.161 616 ...

6 Here are the top four distances thrown by the world's best javelin throwers.

 Steve   0.092 424 km

 Jan   92.336 m

 Mikael 9227.8 cm

 Hans   923 000 mm

 a Round each athlete's distance to the nearest unit.

 b Rank the competitors.

 c Is this a fair way of identifying the best javelin thrower? Explain your answer.

7 **Puzzle**
Georgina cubes a number to give an answer of 300 (correct to the nearest whole number).
What number did she cube (give your answer correct to 2 decimal places)?

Use the power key on your calculator $\boxed{x^y}$

This spread will show you how to:

▶▶ Consolidate and extend mental methods of calculation for addition and subtraction.

▶▶ Consolidate standard column procedures for addition and subtraction of integers and decimals with up to 2 dp.

▶▶ Make and justify estimates and approximations.

KEYWORDS

Addition   Estimate

Subtraction   Partition

Decimal number

You should always see if you can work out additions and subtractions in your head.

Here are two useful methods:

Partitioning

This involves breaking a number into parts that are easier to work with in your head.

$$^-237 - {}^-73 = {}^-237 + 73$$

$$= {}^-237 + 70 + 3$$
$$= {}^-164$$

Compensation

This involves rounding a number up or down and then compensating by adding or subtracting the extra amount.

$$0.84 + {}^-0.69 = 0.84 - 0.69$$

$$= 0.84 - 0.7 + 0.01$$
$$= 0.14 + 0.01$$
$$= 0.15$$

When numbers are too difficult to calculate in your head you should use a written method.

Remember to estimate the answer first.

example

Calculate:

a $33.87 + 165.9 + 23.654$

b $27.5 - {}^-52.3 - 37.27$

a Estimate: $30 + 170 + 20 = 220$

Line up the decimal points:
```
   33.87
  165.9
+  23.654
  223.424
```

$33.87 + 165.9 + 23.654 = 223.424$

b Rewrite the calculation:
$27.5 + 52.3 - 37.27$

Estimate: $30 + 50 - 40 = 40$

Add mentally:
$27.5 + 52.3 = 79.8$

Line up the decimal points:
```
   79.80     Add a zero
  -37.27
   42.53
```

$27.5 - {}^-52.3 - 37.27 = 42.53$

Exercise N3.3

1 Calculate these using a mental or written method.
  a ⁻23 + 27
  b 167 − ⁻49
  c 240 + 160 + ⁻190
  d 47.4 − 11.5
  e 21.7 + 34.8 + 17.2
  f 18.3 − 6.8 + 13.2 − 3.5

2 **a** A diver stands at a height of 17.3 m above the bottom of an empty swimming pool. The swimming pool is filled with water to a depth of 4.48 m. How many metres above the water is the diver?
  b A walker records the climbs and descents on a circular walk. Here are her notes:

Climb	218 m
Descent	148 m
Climb	330 m
Descent	65 m
Climb	146 m
Descent	211 m
Climb	54 m
Descent	324 m

  i How many metres did she climb during her walk?
  ii What was the highest point of the walk? Explain your answer.

3 Calculate these, using an appropriate method:
  a ⁻4.7 + 2.4
  b 3.7 − ⁻2.9 − 1.6
  c 37.17 + 6.7 − 12.26
  d 37.5 + 42.8 − 17.31
  e 1.43 − 16.8 + 14.9
  f 46.3 + 178.9 + ⁻81.26
  g 0.074 kg − 0.098 kg
  h $3\frac{2}{5}$ m − 1.75 m + ⁻0.6 m
  i 1.83 km − ⁻0.98 km
  j 2.07 mg + 4.3 mg + ⁻7 mg + 0.95 mg

4 **a** Steve is 30 cm shorter than Peter. Bernard is 86 mm shorter than Carl. Peter is 1.2 cm taller than Bernard. Peter is 1.84 m tall. What are the heights of Steve, Carl and Bernard?
  b Calculate the perimeter of this paddock in kilometres.

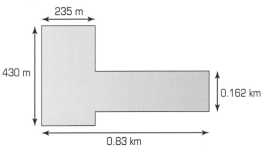

  c The population of Austria in the year 2000 was 8.14 million. The population of Austria in the year 2050 is expected to be 7.094 million. By how much is the population expected to change over the 50 years?

5 What needs to be added or subtracted to change:
  a 5.238 into 5.258
  b 7.611 into 7.618
  c 3.186 into ⁻3.169
  d 12.658 into 12.734

6 **Puzzle**
The difference between two numbers is 7.055. The number exactly halfway between the two numbers is 6.7775.
  a What are the two numbers?
  b Explain the method you used to solve this problem.

7 **Puzzle**
Fill in the boxes with the digits 1 to 9, using each digit only once.

Mental multiplication and division

This spread will show you how to:

▶▶ Consolidate and extend mental methods of multiplication and division.

KEYWORDS

Multiplication   Partition

Division   Factor

You should always see if you can work out multiplications and divisions in your head.

Here are three mental methods:

Using factors

This involves rewriting a number as two of its factors.

example

Calculate: **a** 37×0.04 **b** $525 \div 15$

$$37 \times 0.04 = 37 \times 4 \times 0.01$$
$$= 148 \times 0.01$$
$$37 \times 0.04 = 1.48$$

$$525 \div 15 = 525 \div 5 \div 3$$
$$525 \div 5 = 105$$
$$105 \div 3 = 35$$
$$525 \div 15 = 35$$

Using partitioning

This involves splitting a number into parts and multiplying each part separately.

example

Calculate: **a** 17×1.3 **b** $^-2.6 \times 31$

$$17 = 10 + 7$$
$$17 \times 1.3 = (10 \times 1.3) + (7 \times 1.3)$$
$$= 13 + 9.1$$
$$17 \times 1.3 = 22.1$$

$$31 = 30 + 1$$
$$^-2.6 \times 31 = (^-2.6 \times 30) + (^-2.6 \times 1)$$
$$= ^-78 + ^-2.6$$
$$= ^-78 - 2.6$$
$$^-2.6 \times 31 = ^-80.6$$

Using doubling and halving

This involves doubling one of the numbers and halving the other.

example

Calculate: **a** $^-6.4 \times 2.5$ **b** 22×3.2

$$^-6.4 \times 2.5 = ^-6.4 \times 2.5$$
$$= ^-3.2 \times 5$$
$$^-6.4 \times 2.5 = ^-16$$

$$22 \times 3.2 = 11 \times 6.4$$
$$= (10 \times 6.4) + (1 \times 6.4)$$
$$= 64 + 6.4$$
$$22 \times 3.2 = 70.4$$

Exercise N3.4

1 Calculate:

a 15×12 b 0.4×9
c 0.7×9 d $108 \div 6$
e 20×0.6 f 28×11
g $23 \div 0.1$ h 0.8×18
i 3.2×0.01 j $184 \div 8$
k 0.4×29 l 1.4×40

2 a Hannah swims every 50 metres in 32 seconds. How long will it take her to swim 400 metres?

b Pencils are packaged into boxes of 12. How many boxes will be needed for 312 pencils?

c Each week, the lowest nightly temperature is recorded in Anchorage, Alaska. For the first week the lowest temperature was recorded as $^-12.6$ degrees. Over the next 16 weeks the lowest nightly temperature fell by a further 0.8 degrees each week. What was the lowest nightly temperature for the 17th week?

d Clint pays for 12 bars of chocolate with a £10 note. Each bar of chocolate costs £0.45. How much change does he receive?

3 Puzzle

Copy and complete this multiplication grid:

×		0.7		11
3	0.3			
				55
	1.6		128	

a Write down any strategies you used to solve this problem.

b Use your grid to work out 19.8×24

4 Calculate these. You may need to make some jottings.

a 17×1.1 b $^-1.4 \times 21$
c 3.5×16 d $144 \div 18$
e 39×0.06 f $^-1.6 \div 0.1$
g 18×2.1 h 6.8×29
i $^-19 \times 7$ j $^-435 \div -15$
k 5.4×14 l $^-2.6 \times 2.5$

5 Puzzle

Find the correct digit to go in each of these boxes. Some of the questions have more than one solution.

a $\square 4 \times 1\square = 2\square 4$

b $18 \times \square.6 = \square 4.8$

c $^-\square.4 \times \square 1 = ^-\square 1.4$

d $^-4\square\square \div 15 = ^-\square 8$

6 In these multiplication trails you must choose three numbers, one from each row, which multiply together to give the target number:

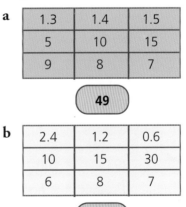

a

1.3	1.4	1.5
5	10	15
9	8	7

49

b

2.4	1.2	0.6
10	15	30
6	8	7

84

7 Find the missing number in each of these calculations. You should try to do all calculations mentally, but you may need to make some jottings.

a $37 \times ? = 1.48$ b $702 \div ? = 18$
c $^-4.6 \times ? = 101.2$ d $? \times 3.5 = ^-94.5$
e $540 \div 12 = ?$ f $8.2 \times ? = 4.92$

This spread will show you how to:

▶▶ Use standard column procedures for multiplication of integers and decimals.

▶▶ Understand where to position the decimal point by considering equivalent calculations.

▶▶ Make and justify estimates and approximations of numbers and calculations.

You can multiply with decimals by considering equivalent calculations.

example

Calculate: **a** 0.5×0.3 **b** 27.4×9.8 **c** 2.93×47

a Working mentally:

$5 \times 3 \quad = \quad 15$

$0.5 \times 3 \quad = \quad 1.5$

$0.5 \times 0.3 \quad = \quad 0.15$

$0.5 = \frac{5}{10} = 5 \div 10$
so $0.5 \times 3 = 5 \times 3 \div 10$
$0.5 \times 0.3 = 5 \times 3 \div 10$
$\quad\quad\quad\quad = 1.5 \div 10$

b Using the grid method:

Estimate $\quad 27.4 \times 9.8 \approx 27 \times 10 = 270$

Partition: $\quad 27.4 \quad = 20 + 7 + 0.4$

$\quad\quad\quad\quad\quad 9.8 \quad = 9 + 0.8$

	20	7	0.4
9	180	63	3.6
0.8	16	5.6	0.32

Add:

$27.4 \times 9.8 \quad = 180 + 63 + 16 + 3.6 + 5.6 + 0.32$

$\quad\quad\quad\quad\quad = 268.52$

$4 \times 8 \quad = 32$
$0.4 \times 8 \quad = 3.2$
$0.4 \times 0.8 = 0.32$

c Using the standard column method:

Estimate: $\quad 2.93 \times 47 \approx 3 \times 50 = 150$

Write an equivalent calculation:

$2.93 = 2\frac{93}{100} = \frac{293}{100} = 293 \div 100$

So $2.93 \times 47 = 293 \times 47 \div 100$

Use the standard column method for the integers:

$$293$$
$$\times \underline{47}$$

$293 \times 40 \quad 11720$

$293 \times 7 \quad \underline{2051}$

$\quad\quad\quad\quad \underline{13771}$

So $\quad 2.93 \times 47 \quad = 293 \times 47 \div 100$

$\quad\quad\quad\quad\quad\quad = 13\,771 \div 100$

$\quad\quad\quad\quad\quad\quad = 137.71$

Exercise N3.5

1 Calculate mentally:
- **a** 1.3×11
- **b** 0.7×7
- **c** 23×15
- **d** 8.2×60
- **e** 3.7×19
- **f** 3.5×12
- **g** 7.3×30
- **h** 9.1×40

2 Investigation

Joanne works out the multiplication:
$$28 \times 41 = 1148$$
She reverses the numbers and multiplies again:
$$82 \times 14 = 1148$$
Find some more pairs of numbers which give the same product when the digits are reversed.
Can you find a way of predicting which pairs of numbers will work?

3 Calculate these, using a mental or written method as appropriate.
- **a** 0.7×0.6
- **b** 4.37×6
- **c** 8.15×7
- **d** 12.3×0.8
- **e** 26.2×3.7
- **f** 16.4×2.5
- **g** 2.46×35
- **h** 19×3.2
- **i** $48 \times £7.28$
- **j** $14.7 \text{ m} \times 8.3 \text{ m}$
- **k** $18.8 \text{ km} \times 5.3$

4
- **a** A length of wood from the timber merchant costs £1.40 a metre. Andy needs 17.3 metres of wood. How much will the wood cost Andy?
- **b** A case of 24 tins of baked beans weighs 9.96 kg. What is the weight of 14 cases?
- **c** A magnifying glass makes objects appear 3.4 times larger than they really are. How long would a worm of length 7.3 cm appear to be in the magnifying glass?

5 Here is some information from a 500g packet of pasta:

Nutritional Information:

Composition	100g provides
Energy	345 kcal
Protein	13.2 g
Carbohydrate	68.5 g
Fat	2.0 g
Fibre	2.9 g

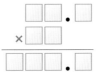

A 100 g serving of dry pasta weighs 250 g when cooked.

Suitable for vegetarians.

ONLY 39 pence

- **a** What is the carbohydrate content of 450 g of dry pasta?
- **b** What is the cost of 15 packets of pasta?
- **c** How much would 68 g of dry pasta weigh when cooked?
- **d** How much fibre is there in 160g of dry pasta?

6 Calculate these products and then rearrange from lowest to highest to reveal a meeting point.

T	33.4×3.5		O	218×5.7
I	114×9.1		E	85.4×3.7
E	2.3×67.6		T	51.8×9.7
N	60×22.5		I	0.78×64
R	27.6×8.1		C	3.9×95.7
N	16.3×3.4		S	3.97×75

7 Puzzle

Copy and complete this multiplication using each of the digits 1 to 9 only once.

8 Calculate the area of this sandpit. Give your answer to the nearest m^2.

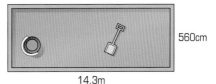

560cm

14.3m

This spread will show you how to:
▶▶ Use standard column procedures for division of integers and decimals.
▶▶ Make and justify estimates and approximations of calculations.
▶▶ Check a result by considering whether it is of the right order of magnitude and by working the problem backwards.
▶▶ Interpret the display on a calculator.

KEYWORDS
Division   Estimate
Remainder
Order of magnitude

You can use repeated subtraction to divide numbers.
You can check an answer to a division by multiplying, and by considering whether it is the right order of magnitude.

example

Calculate: **a** $39.6 \div 18$ **b** $73.9 \div 13$ (to 1 dp)

a Estimate: $39.6 \div 18 \approx 40 \div 20 = 2$

$$18)39.6$$
$$\underline{-36.0} \qquad 18 \times 2 = 36$$
$$3.6$$
$$\underline{-3.6} \qquad 18 \times 0.2 = 3.6$$
$$0.0$$
$$39.6 \div 18 = 2.2$$

Think
$18 \times 0.1 = 1.8$
$18 \times 0.2 = 3.6$

Check by multiplying: $18 \times 2.2 = 39.6$

The answer in **b** is not the same because you are using a rounded number.

b Estimate: $73.9 \div 13 \approx 70 \div 14 = 5$

$$13)73.90$$
$$\underline{-65.00} \qquad 13 \times 5 = 65$$
$$8.90$$
$$\underline{-7.80} \qquad 13 \times 0.6 = 7.8$$
$$1.10$$
$$\underline{-1.04} \qquad 13 \times 0.08 = 1.04$$
$$0.06$$

$73.9 \div 13 = 5.68$ remainder 0.06
$73.9 \div 13 = 5.7$ (1 dp)

Check by multiplying: $5.7 \times 13 = 74.1$

You can express a remainder as a number, fraction or decimal.
The way you write the remainder depends on the question.

example

243 pence is shared between 7 people.
How much money will each person receive.

Using a calculator: $243 \div 7 = 34.714\ 285 \ldots$
Each person will receive 34.71 pence (to 2 dp).
You cannot receive 0.71 of a penny, so this is not a sensible answer.
Write the remainder as a whole number:
$243 \div 7 = 34$ remainder 5
Each person will receive 34 pence (with 5 pence left over).

To change the remainder to a whole number, multiply the decimal part by the divisor:
$0.\dot{7}1\dot{4}\ \dot{2}8\dot{5} \times 7 = 4.999 \ldots$
≈ 5

Exercise N3.6

1 Copy and complete this shopping bill:

Item	Cost per item	Number of items	Total cost
Bread (loaves)	£0.42	5	
Coffee (500 g)		6	£22.44
Pasta (500 g)		7	£ 4.48
Margarine (500 g)		3	£ 5.34
Wine (0.75 litre)		8	£28.72
Tinned tomatoes		9	£ 2.07
Cheese (200 g)	£2.62	3	
		Total	

2 Calculate (giving your answers to 1 dp as appropriate):
  a $113.9 \div 17$ **b** $88.3 \div 14$ **c** $181.7 \div 23$ **d** $243.2 \div 64$
  e $200 \div 38$ **f** $155.1 \div 33$ **g** $212.8 \text{ m} \div 28$ **h** $109.2 \text{ kg} \div 15$

3 **a** How much money is left over when 12 friends share £37?
  b A car transporter weighs 12.6 tonnes when empty and 23.4 tonnes
    when loaded with 12 cars. What is the weight of one car?
  c 44 litres of petrol costs £31.60. What is the cost of one litre of petrol
    (to the nearest tenth of a penny)?
  d The total weight of the 11 players at Gowing Town FC is 873.5 kg.
    What is the mean weight of a footballer (to 1 dp)?

4 Solve each of these problems, deciding how to express the answer.
  a The maximum load for a forklift truck is 475 kg. A box weighs 32 kg.
    How many boxes can the truck lift?
  b The area of a square flowerbed is 6 m². What is the length of the side of the flowerbed?
  c Is 10 000 minutes longer or shorter than a week?
    By how much? Explain your answer.

5 **Investigation**
  Follow the instructions in the flow chart.
  Investigate what happens for different start numbers.
  Use your calculator to investigate further, by
  changing some of the instructions.
  Write down anything you notice.

6 **Puzzle**
  Baked beans come in three different-sized tins.
    100 g costs 15p
    250 g costs 37p
    450 g costs 67p
  Which size of tin is the best value for money?
  Explain your answer.

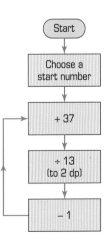

This spread will show you how to:

▶▶ Reduce a ratio to its simplest form, including a ratio expressed in different units, recognising links with fraction notation.

▶▶ Divide a quantity into two or more parts in a given ratio.

▶▶ Use the unitary method to solve problems.

KEYWORDS

Ratio   Fraction
Proportion   Compare

▶ You use a ratio to divide something into different-sized parts.

example

Mike, Betty and Glyn shared their lottery winnings of £7000 in the ratio 3 : 5 : 6. How much did each person receive?

The £7000 has to be split into $3 + 5 + 6 = 14$ equal parts

Each part = £7000 ÷ 14 = £500

Mike received 3 parts   = 3 × £500 = £1500

Betty received 5 parts   = 5 × £500 = £2500

Glyn received 6 parts   = 6 × £500 = £3000

You can see the proportions:
Mike received $\frac{3}{14}$
Betty received $\frac{5}{14}$
Glyn received $\frac{6}{14}$

▶ You can use a ratio to compare the sizes of two or more quantities.

Axel weighs 4 kg.

Morrissey weighs 10 kg.

Axel's weight : Morrissey's weight

     = 4 kg : 10 kg

     = 4 : 10

     = 2 : 5

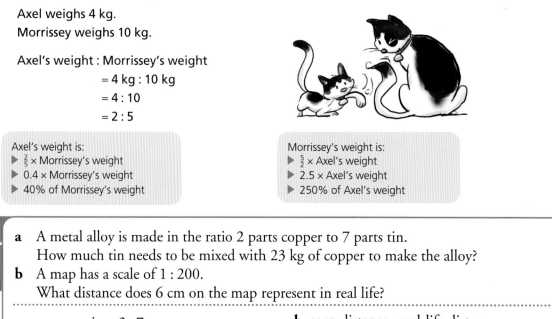

Axel's weight is:
▶ $\frac{2}{5}$ × Morrissey's weight
▶ 0.4 × Morrissey's weight
▶ 40% of Morrissey's weight

Morrissey's weight is:
▶ $\frac{5}{2}$ × Axel's weight
▶ 2.5 × Axel's weight
▶ 250% of Axel's weight

example

a A metal alloy is made in the ratio 2 parts copper to 7 parts tin.
How much tin needs to be mixed with 23 kg of copper to make the alloy?

b A map has a scale of 1 : 200.
What distance does 6 cm on the map represent in real life?

a   copper : tin = 2 : 7

       tin = $\frac{7}{2}$ × copper

         = 3.5 × 23 kg

         = 80.5 kg

b   map distance : real-life distance

         = 1 : 200

    real life = $\frac{200}{1}$ × map

         = 200 × 6 cm

         = 1200 cm

Exercise N3.7

1 Write these ratios in their simplest form:
  a $12:20$ **b** $8:36$ **c** $150:500$
  d $13:29$ **e** $17:357$

2 **a** In a Year 8 tutor group there are 6 boys for every 5 girls. There are 18 boys in the tutor group. How many girls are in the tutor group?
  b There are 4 cream cakes for every 7 plain cakes in a shop. There are 297 cakes in the shop. How many cream cakes are there in the shop?

3 Express these ratios in their simplest form.
  a A recipe requires 400 g of flour for every $\frac{1}{2}$ kg of butter. What is the ratio of flour to butter?
  b Albert has £350, Del has £300 and Rodney has £150.
    What is the ratio of Albert's : Del's : Rodney's money?
  c Gavin is 1.82 m tall. James is 168 cm tall. What is the ratio of James' height to Gavin's height?
  d Alan earns £1800 a month. He spends $\frac{1}{5}$ of his earnings on his mortgage, $\frac{1}{8}$ on bills and $\frac{1}{10}$ on groceries. What is the ratio of money spent on bills to money spent on groceries?

4 **a** A 720 cm length of wood is divided into two pieces in the ratio $11:7$. How long is each of the two pieces?
  b A garage bill of £676.80 is shared between Sam, Steve and Sandra in the ratio $3:4:5$. How much does each person have to pay?
  c A pie chart is drawn to represent the ratio of cars : vans : lorries at a motorway service station. The ratio of cars : vans : lorries was $11:5:2$. How big is the sector angle representing the number of cars on the pie chart?

5 Copy and complete this diagram adding at least 3 more statements that are true.

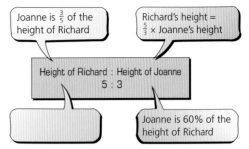

6 **a** An alloy is made from iron and copper in the ratio $5:4$. How much iron is needed to mix with 68 kg of copper?
  b Paula and Caroline are given pocket money each week in the ratio $7:13$. If Paula is given £10.50 a week, how much pocket money is Caroline given?
  c The scale of a map is $1:10\,000$. How long is a distance of 8 km in real life shown on the map?
  d The ratio of KS3 pupils to KS4 pupils in a school is $3:2$. There are 714 KS3 pupils. How many KS4 pupils are there at the school?

7 **Investigation**
Here are some common map scales:
$1:2\,000\,000$; $1:2000$; $1:500$.
  a What distance does 4 cm on each map represent in real life?
  b Investigate some other distances. What kind of map should each scale be used for? Explain and justify your answer.

8 **Investigation**
£1500 is split between three sisters.
Amanda has less than $\frac{1}{3}$ of the total.
Jane has more than $\frac{1}{3}$ of Louises share.
Louise has less than 160% of Amanda's share.
Split the money using a whole number ratio so that it satisfies all of these conditions.

This spread will help you to:

▶▶ Consolidate understanding of the relationship between ratio and proportion.

KEYWORDS

Ratio   Scale factor

Inverse

▶ A ratio written as a multiplier is called a **scale factor**.

A scale factor tells you how many times bigger one number is compared with another.

example

A photograph is 10 inches wide.
An enlargement of the same photograph is 16 inches wide.
How many times wider is the enlargement?

Scale factor $= \dfrac{\text{Width of enlargement}}{\text{Width of original}}$

$= \dfrac{16}{10} = \dfrac{8}{5}$

$= 1.6$

The enlargement is 1.6 times wider.

So you can say:
Width of enlargement $= 1.6 \times$ width of original
$= \frac{8}{5} \times$ width of original
$= 160\%$ of width of original
Width of enlargement : width of original $= 8 : 5$

To find the width of the original you use the inverse scale factor.

▶ You can divide to find the inverse ... or you can multiply ...

example

Porky pig is $\frac{9}{4}$ times as heavy as Podgy pig.
Write two more ways to compare the weight of the pigs.

Weight of Porky pig $= \frac{9}{4} \times$ weight of Podgy pig
Weight of Podgy pig $= \frac{4}{9} \times$ weight of Porky pig
Porky's weight : Podgy's weight $= 9 : 4$

Exercise N3.8

1 Calculate:
  a $\frac{3}{5}$ of £220 **b** 45% of $40 **c** $\frac{4}{7} \times 364$ m **d** $\frac{7}{4} \times 364$ m
  e 48 kg × 2.5 **f** 37.2 × 0.1 **g** 15.4 × 2.7 **h** 145% of 23 litres

2 **Investigation**
  a Find three different ways to get from 8 to 20 using only multiplication and division.
  b How can you get from 8 to 20 using a single multiplication?
  c Find three different ways of getting from 20 to 8 using only multiplication and division.
    What do you notice?
  d How can you get from 20 to 8 using a single multiplication?
  e Investigate finding a single multiplication to go from one number to another for different pairs of numbers.

3 Copy and complete this table comparing prices in 1985 and 2003:

Item	Price in 1985	Price in 2003	Price increase
Milk (1 pint)	19 pence	38 pence	× 2
Colour TV set	£320		× 1.25
Mars bar	16 pence	40 pence	
Calculator	£6.40	£3.84	

4 **a** Shagufta scored $3\frac{1}{2}$ times as many marks as Marcus in her French test.
    Marcus scored 24 marks. How many marks did Shagufta score?
  b The maximum speed of the Hawker FD3Xi jet fighter is 1.63 times the speed of sound.
    The speed of sound is about 1200 kph. What is the maximum speed of the jet?
  c In a library book $\frac{2}{5}$ of the pages have a photograph on them.
    There are 465 pages in the book.
    How many pages have photographs on them?

5 What fraction of:
  a 180 is 40 **b** 3 hours is 45 minutes **c** 2 km is 1200 metres?

6 **a** Chandler spends $2\frac{1}{4}$ times as much as he saves.
    i If he spends £771.39 in a month, how much does he save?
    ii What percentage of his money does he spend each month?
  b Frodo is $1\frac{3}{8}$ times as tall as Bilbo.
    i What percentage of Frodo's height is Bilbo's height?
    ii What is the ratio of Frodo's height : Bilbo's height?

7 **Puzzle**
From January 2002 to December 2002 the value of a share in the company DOTCOM fell by 85%. The price in December 2002 was 12.6 pence.
What was the value of a share in January 2002?

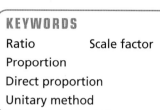

This spread will help you to:

⏵⏵ Consolidate the relationship between ratio and proportion.

⏵⏵ Use the unitary method to solve problems involving direct proportion.

KEYWORDS

Ratio   Scale factor

Proportion

Direct proportion

Unitary method

▶ Numbers or quantities are in proportion when the ratio of corresponding values is always the same.

The table shows some conversion values of miles and kilometres. In each case the ratio of each pair of values is the same.

Number of miles	Number of km	Ratio
25	40	$25 : 40 = 5 : 8$
10	16	$10 : 16 = 5 : 8$
12	19.2	$12 : 19.2 = 5 : 8$
2.5	4	$2.5 : 4 = 5 : 8$

The number of miles is **proportional** to the number of km.

If one of the quantities changes, the other quantity changes in the same proportion. If you double the number of miles, you double the number of kilometres.

▶ You can use direct proportion to solve problems.

There are three common methods.

example

£8.00 is worth 14.40 U.S. dollars. What is the value of £5.00?

Unitary method

Find the value of £1.00 in US dollars

$\times \frac{1}{8}$ (£8.00 is worth $14.40) $\times \frac{1}{8}$

$\times 5$ (£1.00 is worth $1.80) $\times 5$

£5.00 is worth $9.00

Scaling method

The proportion by which each number has changed $= \frac{£5.00}{£8.00} = \frac{5}{8}$

$\times \frac{5}{8}$ (£8.00 is worth $14.40) $\times \frac{5}{8}$

£5.00 is worth $9.00

Scale factor method

Scale factor (ratio) $= \dfrac{\text{Number of US dollars}}{\text{Number of pounds}} = \dfrac{14.40}{8.00} = 1.8$

£5.00 is worth $5 \times 1.8 = $9.00

$\div 1.8$

$\$$ $£$

$\times 1.8$

Exercise N3.9

1 Solve these problems using mental or written methods.
   a 3 kg of potatoes cost 80 pence. How much do 12 kg cost?
   b 15 chocolate bars cost £6.30. How much do 3 cost?
   c 0.7 kg of flour costs £0.68. How much do 7 kg cost?
   d 3 CDs cost £35.97. How much do 6 CDs cost?

2 Copy and complete this table showing the length of various films
in their original form and then as the 'director's cut'.

Film	Original length (mins)	Director's cut (mins)	Scale factor
Geometry Day	140	154	
The E-Quation	80	108	
Cumulative Frequency	115		$\times 1.2$
The Fraction Conversion		140	$\times \frac{7}{6}$

3 Identify which of these sets of numbers are in direct proportion
by calculating the scale factors. Explain and justify your answers.

a

a	b
3	4.2
0.7	0.98
8	11.2
15	21
28	39.2

b

x	y
7	29.4
3	13.4
0.4	3
30	121.4
4.5	19.4

c

l	m
6	36
1.5	2.25
8.5	72.25
14	196
0.7	0.49

4 To make 9 litres of green paint you mix 4 litres of blue paint with
5 litres of yellow paint.
   a Explain how you would make these quantities of green paint:
     i 18 litres   **ii** 4.5 litres
     iii 0.9 litres   **iv** 1 litre
     v 13 litres   **vi** 25.6 litres
   b What percentage of the green paint is blue?

5 Solve each of these problems, clearly showing your method.
   a 15 litres of oil cost £19.80. How much would 20 litres of oil cost?
   b On average, Mylene's car travels 211 miles using 5 gallons of petrol.
     How many miles will it travel with 9 gallons of petrol?
   c 12 pizzas cost £18. How much would 4 pizzas cost?
   d A distance of 31 m is represented by 12.4 cm on my scale drawing.
     How many metres are represented by a distance of 8 cm on my drawing?
   e A 450 g packet of rice costs £1.71. How much per 100 g of rice is this?

This spread will show you how to:
▶▶ Know that if two 2-D shapes are congruent, corresponding sides and angles are equal.
▶▶ Congruent shapes are exactly the same shape and size.

These shapes are ...

... the same shape and size

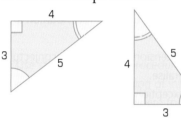

... congruent.

... the same shape but different sizes.

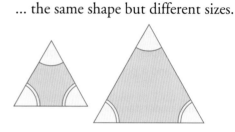

... not congruent.

Triangles ABC and PQR are congruent.

AB = PR
AB corresponds to PR.

BC = RQ
BC corresponds to RQ.

AC = PQ
AC corresponds to PQ.

$\hat{A} = \hat{P}$
A corresponds to P.

$\hat{B} = \hat{R}$
B corresponds to R.

$\hat{C} = \hat{Q}$
C corresponds to Q.

▶ When shapes are congruent:
  ▶ corresponding sides are equal in length, and
  ▶ corresponding angles are equal.

Describe pairs of shapes that are congruent.
Justify your answers.

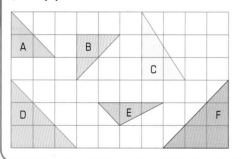

A and B are congruent.
Angles and sides are equal.

D and F are congruent.
Angles and sides are equal.

Exercise S3.1

1 Match these shapes into congruent pairs. Which is the odd one out?

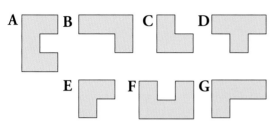

2 a Explain why these triangles are congruent:

b These triangles are congruent:

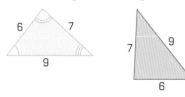

Label the angles on the second triangle which correspond to the angles on the first triangle.

c Which of these shapes are congruent? Give reasons for your answers.

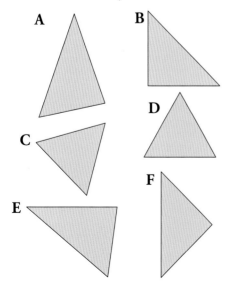

3 Investigation

You need square dotty paper.
You can divide this 4 × 4 pinboard into two congruent halves by joining two dots together. For example:

a Investigate the number of ways you can do this.

b Investigate further with a 5 × 5 pinboard and a 6 × 6 pinboard.

4 Investigation

You need square dotty paper.
On a 3 × 3 grid, draw a triangle.
Label it A.

a Draw as many different congruent triangles as you can.
For example, starting with:

Here are two possible congruent triangles:

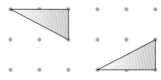

b Repeat for other different starting triangles.

c Repeat for different quadrilaterals.

d Which starting shape produces the most different congruent shapes?

e Investigate whether your results would be the same using other sized grids.

This spread will show you how to:

▶▶ Transform 2-D shapes.

KEYWORDS
Translation   Object
Rotation   Image
Reflection   Map

You can describe the movement or transformation of a shape on a grid. The movement is called a mapping.

All the shapes are exactly the same shape and size so they are all congruent.

▶ A translation slides the object.
  You specify:
   ▶ the distance left or right, and then
   ▶ the distance up or down.

The transformation that maps A onto C is a translation of 2 units across and 1 unit down.

You can write this using vector notation: $\begin{pmatrix} 2 \\ -1 \end{pmatrix}$ means 2 across and 1 down.

▶ A reflection flips the object over.
  You specify the mirror line.

The transformation that maps B onto F is a reflection in the x-axis.
The transformation that maps A onto E is a reflection in the line $x = 1$.

▶ A rotation turns the object.
  You specify:
   ▶ the centre of rotation   ▶ the angle of turn
   ▶ the direction of turn.

The transformation that maps A onto D is a rotation about (0, 0) through 90° clockwise.
The transformation that maps C onto B is a rotation about (3, 2) through 180°.

Exercise S3.2

1   Describe the translation given by each of these vectors.

   a $\begin{pmatrix} 4 \\ 3 \end{pmatrix}$ b $\begin{pmatrix} 3 \\ 4 \end{pmatrix}$ c $\begin{pmatrix} -2 \\ 1 \end{pmatrix}$ d $\begin{pmatrix} 1 \\ -4 \end{pmatrix}$ e $\begin{pmatrix} -3 \\ -2 \end{pmatrix}$

2   a   Write the vectors that translate:
       i   A to C   ii A to D
       iii C to B   iv G to H
       v   H to G   vi I to G

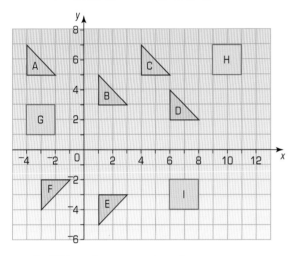

   b   What do you notice about your answers to parts **iv** and **v**?
   c   Can B be translated to E or F? Explain your answer.

3   Describe the transformation that will map:
   a   A onto B   b   B onto C
   c   D onto C   d   A onto E
   e   C onto E   f   F onto A

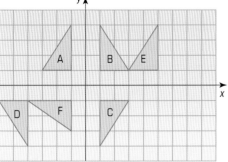

4   Copy this diagram.
   a   Reflect A in the y-axis. Label it A′.
   b   Reflect A in the x-axis. Label it A″.
   c   Reflect A′ in the x-axis. Label it A‴.
   d   Describe the reflections that map:
       i   A onto A‴   ii   A‴ onto A
       Explain what you notice.

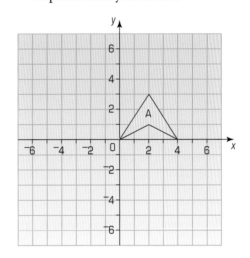

5   Copy this diagram. Label the rectangle R.
   a   Rotate R 90° clockwise about the origin. Label the image R′.
   b   Rotate R′ 180° about the origin. Label the image R″.
   c   What rotation will map:
       i   R″ onto R   ii   R onto R″?
       Explain what you notice.

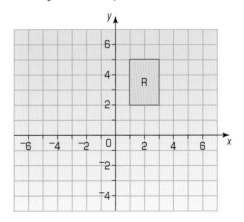

Combining transformations

KEYWORDS

Rotation   Image

Reflection

Translation

This spread will show you how to:

▶▶ Transform 2-D shapes by combining simple combinations of rotations, reflections and translations.

You can combine different transformations together.
You can often describe the result as a single transformation.

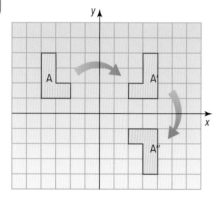

A is reflected in the *y*-axis to produce A'.

A' is reflected in the *x*-axis to produce A".

The single transformation that maps A onto A" is a rotation about (0, 0) through 180°.

example

On a grid from ⁻6 to 6 on both axes:

a Draw a triangle, T, with vertices (⁻1, 1), (⁻1, 4), (⁻3, 1).

b Translate the triangle using the vector $\begin{pmatrix} 6 \\ 0 \end{pmatrix}$.
Label this triangle T'.

c Rotate T' about (3, 1) through 180°.
Label this triangle T".

d Describe the single transformation that maps T onto T".

a–c

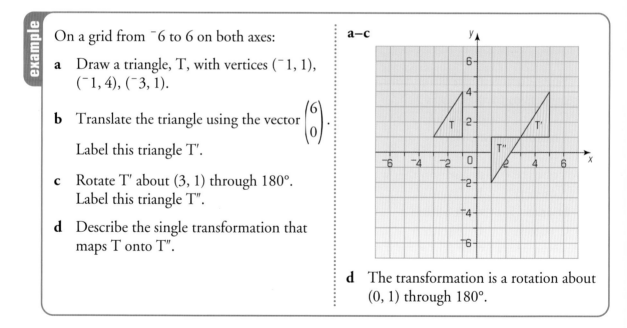

d The transformation is a rotation about (0, 1) through 180°.

The notation A, A' and A" can be confusing.
Read the question carefully and always label shapes immediately.

Exercise S3.3

1 Here are four shapes on a grid:

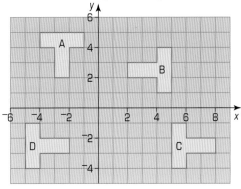

a Find a single transformation that will map:
    i A onto B   **ii** B onto D
    iii A onto D   **iv** D onto C
b Find a combination of two transformations that will map:
    i A onto D   **ii** B onto C
    iii C onto B   **iv** D onto C

2 On a copy of this grid:

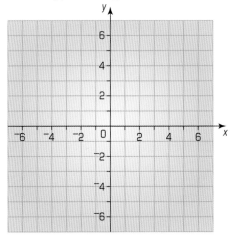

a Draw the triangle with vertices $(^-2, 1)$, $(^-2, 3)$, $(0, 2)$. Label it T.
b Translate the triangle by the vector $\begin{pmatrix} ^-2 \\ ^-3 \end{pmatrix}$. Label the image T'.
c Reflect T' in the y-axis. Label it T".
d What single transformation will map T onto T"?

3 On a copy of the grid in question **2**:
a Draw shape A with vertices $(^-2, 1)$, $(^-4, 3)$, $(^-4, 5)$.
b Reflect A in the x-axis. Label the image A'.
c Reflect A' in the y-axis. Label the image A".
d What single transformation maps A onto A"?

4 Is a rotation, centre $(0, 0)$, through 180° always the same as a reflection in the y-axis followed by a reflection in the x-axis? Justify your answer. You can use this diagram to help you.

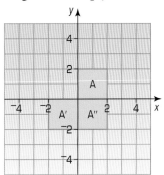

5 What single transformation is equivalent to a reflection in the x-axis followed by a reflection in the y-axis? Justify your answer.

6 **Investigation**
The diagram shows a triangle, T, and two parallel lines, A and B.

▶ Copy the diagram
▶ Reflect T in A. Label it T'
▶ Reflect T' in B. Label it T"
Investigate and explain the relationship:
a the distance between A and B.
b the distance between T" and T.

Describing symmetries

This spread will show you how to:

 Identify all the symmetries of 2-D shapes.

KEYWORDS
Polygon   Regular
Reflection symmetry
Rotational symmetry
Order of rotational symmetry

A polygon is a shape with straight sides.

A polygon may have:

reflection symmetry

rotational symmetry

both

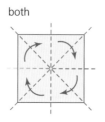

neither

A shape has reflection symmetry if it folds onto itself.

A shape has rotational symmetry if it turns onto itself.

You describe the symmetry of a shape by the number of lines of symmetry and the order of rotational symmetry.

> ► The order of rotational symmetry is the number of times a
>   shape turns onto itself in a complete turn.

The order of rotational symmetry of a parallelogram is 2:

It is back to the original position.

> ► A regular polygon with *n* sides has:
>   ► *n* sides equal in length and
>   ► *n* angles equal.
>   ► *n* lines of symmetry.
>   ► Order of rotational symmetry *n*.

A regular pentagon has 5 equal angles and 5 equal sides.
It has 5 lines of symmetry:

The order of rotational symmetry is 5:

Exercise S3.4

1 For each of these triangles:
- **i** right-angled isosceles (RI)
- **ii** scalene (S)
- **iii** right-angled (R)
- **iv** equilateral (E)
- **v** isosceles (I)

a Write down the number of lines of symmetry.

b Write down the order of rotational symmetry.

c Copy and complete this table inserting the letters RI, S, R, E, I as appropriate:

		Lines of symmetry			
		0	1	2	3
Order of rotational symmetry	0				
	1				
	2				
	3				

2 What is the name of a regular triangle? Explain your answer.

3 Copy and complete this table to classify these quadrilaterals:
- ▶ kite (K)
- ▶ rhombus (R)
- ▶ parallelogram (P)
- ▶ trapezium (T)
- ▶ square (S)
- ▶ isosceles trapezium (IT)

		Lines of symmetry				
		0	1	2	3	4
Order of rotational symmetry	0					
	1					
	2					
	3					
	4					

4 What is the name of a regular quadrilateral? Explain your answer.

5 **Investigation**
Make two copies of this triangle:

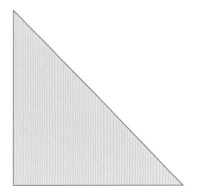

Fit the triangles together along equal sides to make as many different polygons as you can find.
Describe the properties of each polygon you make.
Explain how you know it has the properties you describe.
Investigate using a different starting triangle.

6 **Investigation**
On a 4 × 4 grid or pinboard, draw as many different polygons as you can.

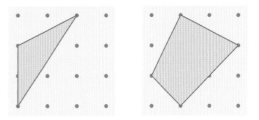

Draw each polygon on a new grid and then write down:
- **a** how many sides it has
- **b** its symmetry properties
- **c** whether it is regular or not.

This spread will show you how to:
- ▶▶ Understand and use the language and notation associated with enlargement.
- ▶▶ Consolidate understanding of the relationship between ratio and proportion.
- ▶▶ Reduce a ratio to its simplest form, including a ratio expressed in different units.

KEYWORDS
Enlargement   Image
Scale factor   Ratio
Object

This photo has been enlarged:

The length of the original is 10 cm.

The corresponding length of the enlargement is 20 cm.

The original length has been multiplied by 2.
The multiplier is the scale factor.

All the lengths of the original photo have been multiplied by 2.

This is an enlargement of scale factor 2.

You can write a scale factor as a fraction or as a ratio:

▶ Scale factor $= \dfrac{\text{Length of enlargement}}{\text{Corresponding length of original}} = \dfrac{2}{1}$

▶ Scale factor = Length of enlargement : Corresponding length of original
$= 2 : 1$

You can project an object to enlarge it.
All the distances from the projector are multiplied.

This is the original object.

This image is an enlargement scale factor 3.

Exercise S3.5

1 Copy these shapes onto squared paper. Enlarge each shape by a scale factor of 2.

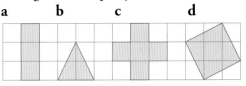

a   b   c   d

2 **a** Copy this shape onto squared paper.
    b Enlarge the shape by a scale factor of 3.

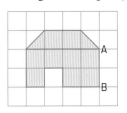

3 Write down the length of the side corresponding to the side AB in question **2** following an enlargement of a scale factor of 4.

4 Find the scale factor of these enlargements:

a

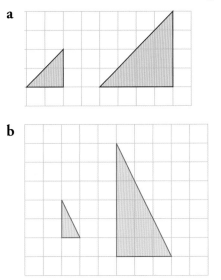

b

c

5 A model car is built using a scale of 1 : 5. Copy and complete this table:

Measurement	Car	Model
Length	250 cm	
Width		26 cm
Number of wheels		5
Length of wipers	80 cm	
Number of seats	4	

6 A model aeroplane is built using a scale of 1 : 20.
Copy and complete this table:

Measurement	Plane	Model
Length	35 m	
Width		1.2 m
Number of seats	127	
Number of engines		2

7 A slide, which is 15 cm away from a projector is projected onto a wall.

The wall is 3 m away from the projector.
a What is the scale factor of the enlargement?
Give your answer in its simplest form.
b The length of the slide is 3 cm.
What is the length of the projection?
c The width of the projection is 2.4 m.
What is the width of the slide?

This spread will show you how to:
- ▶▶ Understand and use the language and notation of enlargement.
- ▶▶ Enlarge 2-D shapes, given a centre of enlargement and a positive whole number scale factor.

KEYWORDS

Centre of enlargement
Scale factor
Vertices

You can enlarge a shape on a grid.
The position of the image depends on the scale factor and the centre.

Here are two different enlargements of a shape by a scale factor of 2:

This enlargement has centre (0, 0).

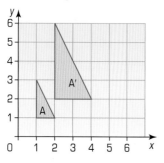

This enlargement has centre (1, 1).

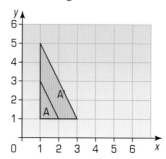

▶ To specify an enlargement you need to give:
  - ▶ the scale factor and
  - ▶ the centre of enlargement.

Lines joining corresponding vertices meet at the centre of enlargement.

example

Describe this enlargement.

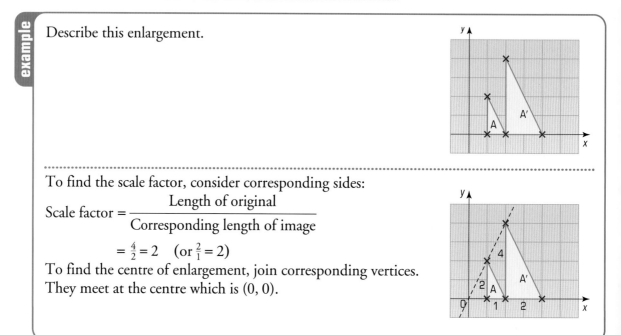

To find the scale factor, consider corresponding sides:

$$\text{Scale factor} = \frac{\text{Length of original}}{\text{Corresponding length of image}}$$

$$= \frac{4}{2} = 2 \quad (\text{or } \frac{2}{1} = 2)$$

To find the centre of enlargement, join corresponding vertices.
They meet at the centre which is (0, 0).

Exercise S3.6

1 Copy this diagram.

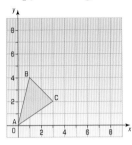

a Enlarge the shape ABC using scale factor 3, centre (0, 0).
Label the enlargement A′B′C′.

b Copy and complete the table of coordinates:

Object	Image
A (0, 0)	A′ (,)
B (1, 4)	B′ (,)
C (3, 2)	C′ (,)

c What do you notice about the coordinates of the image?

d Copy and complete the ratio:
Coordinate of object : Coordinate of image
= :

2 You need four copies of this grid.

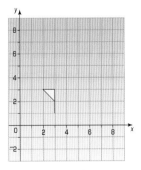

On a new copy of the grid, enlarge the shape using:
a Scale factor 2, centre (0, 0)
b Scale factor 3, centre (1, 2)
c Scale factor 3, centre (2, 0)
d Scale factor 2, centre (6, 2)

3 **a** Find the scale factor of this enlargement.

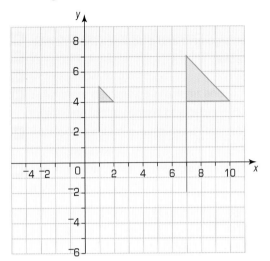

b Find the centre of enlargement.
c Describe the enlargement fully.

4 The diagram shows an image and the centre of enlargement.

The scale factor of the enlargement is 4.

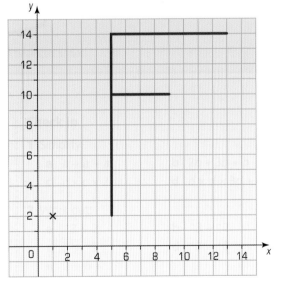

Find the coordinates of the object.

This spread will show you how to:
▶▶ Begin to distinguish between the different roles played by letter symbols.
▶▶ Construct and solve linear equations using inverse operations.

KEYWORDS
Function machine
Equation   Horizontal
Inverse   Vertical

Equations can become quite complicated, so it is important to develop a strategy to help solve them.

One good strategy is to use inverse operations – you unpick the equation to find the unknown value.

Using function machines

The input value is unknown:

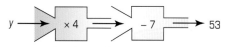

You move backwards, using the inverse operations:

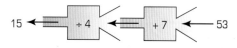

The input is 15.

Using equations

The equation is:

$$4y - 7 = 53$$

To unpick the equation you:

+7: $4y = 53 + 7 = 60$
÷4: $y = 60 \div 4 = 15$
        $y = 15$

---

example

In a Hot Cross, each line adds to the same value.
Form two equations for this Hot Cross and solve them to find r and s.

line total = 44

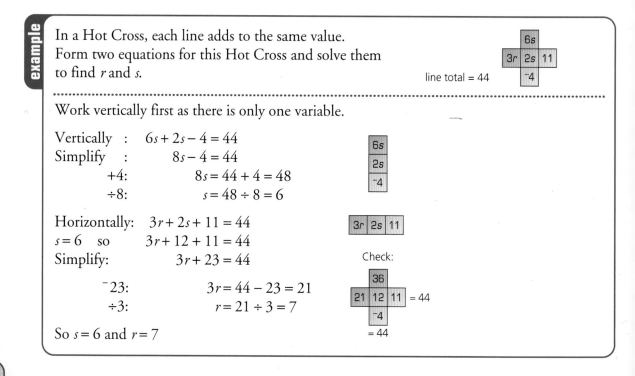

Work vertically first as there is only one variable.

Vertically   : $6s + 2s - 4 = 44$
Simplify   : $8s - 4 = 44$
     +4: $8s = 44 + 4 = 48$
     ÷8: $s = 48 \div 8 = 6$

Horizontally: $3r + 2s + 11 = 44$
$s = 6$ so $3r + 12 + 11 = 44$
Simplify: $3r + 23 = 44$

     ⁻23: $3r = 44 - 23 = 21$
     ÷3: $r = 21 \div 3 = 7$

So $s = 6$ and $r = 7$

Check:

= 44

Exercise A4.1

1 Find the output from these function machines:

a

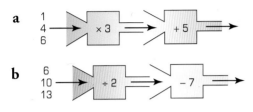

1
4
6
→ ×3 → +5 →

b

6
10
13
→ ÷2 → −7 →

2 Use the inverse method to find the input for these function machines:

a

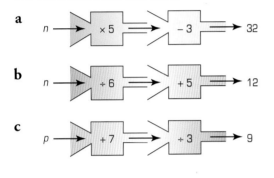

n → ×5 → −3 → 32

b

n → ÷6 → +5 → 12

c

p → +7 → ÷3 → 9

3 Match each equation with its function machine.

a $3x + 5 = 17$ b $5x + 7 = 3$

c $\dfrac{x}{5} + 3 = 17$ d $\dfrac{x}{3} - 17 = 5$

e $5(x + 3) = 30$

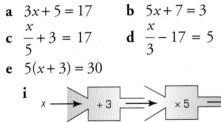

i x → +3 → ×5 → 30

ii x → ÷5 → +3 → 17

iii x → ×3 → +5 → 17

iv x → ×5 → +7 → 3

v x → ÷3 → −17 → 5

Solve each equation.

4 Construct two equations from each Hot Cross. Choose which one to solve first, and work out the value for each letter.

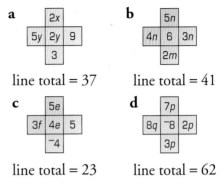

a

	2x	
5y	2y	9
	3	

line total = 37

b

	5n	
4n	6	3n
	2m	

line total = 41

c

	5e	
3f	4e	5
	−4	

line total = 23

d

	7p	
8q	−8	2p
	3p	

line total = 62

5 In a magic square, each horizontal, vertical and diagonal line of three items adds up to the same total.

a Check if these are magic squares.

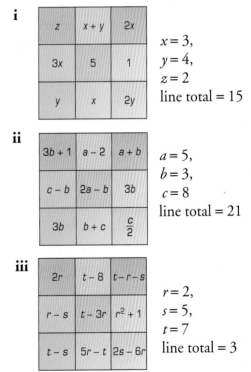

i

z	$x + y$	$2x$
$3x$	5	1
y	x	$2y$

$x = 3$,
$y = 4$,
$z = 2$
line total = 15

ii

$3b + 1$	$a - 2$	$a + b$
$c - b$	$2a - b$	$3b$
$3b$	$b + c$	$\dfrac{c}{2}$

$a = 5$,
$b = 3$,
$c = 8$
line total = 21

iii

$2r$	$t - 8$	$t - r - s$
$r - s$	$t - 3r$	$r^2 + 1$
$t - s$	$5r - t$	$2s - 6r$

$r = 2$,
$s = 5$,
$t = 7$
line total = 3

b In the square which is not magic, change one of the expressions to make it a magic square.

Simplifying expressions

This spread will show you how to:
- Simplify or transform linear expressions by collecting like terms.
- Construct and solve linear equations.

▶ An expression is a collection of letter and number terms.

There are three types of term in this expression:

Collect the like terms:

The simplified expression is:

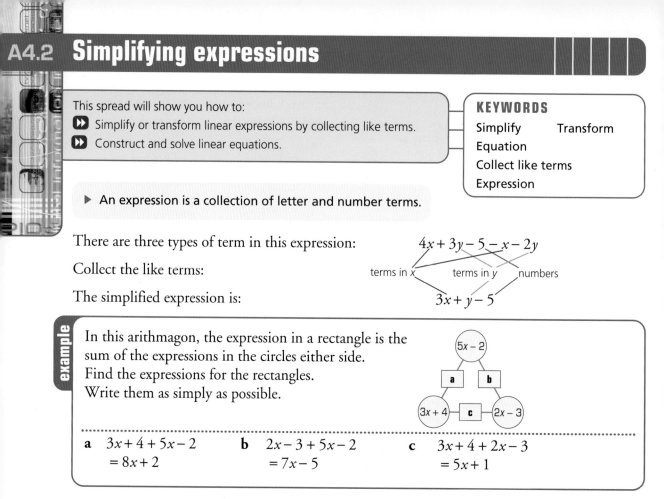

$$4x + 3y - 5 - x - 2y$$

terms in x terms in y numbers

$$3x + y - 5$$

example

In this arithmagon, the expression in a rectangle is the sum of the expressions in the circles either side.
Find the expressions for the rectangles.
Write them as simply as possible.

a $3x + 4 + 5x - 2$
    $= 8x + 2$

b $2x - 3 + 5x - 2$
    $= 7x - 5$

c $3x + 4 + 2x - 3$
    $= 5x + 1$

You can find the missing numbers in this addition grid by adding on:

$7 + ? = 13$
$7 + 6 = 13$

$15 + 6 = 21$

$7 + 13 = 20$

$15 + ? = 28$
$15 + 13 = 28$

The completed grid is:

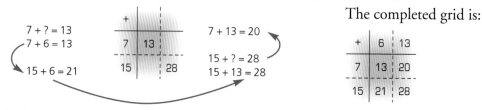

When you have number and letter terms it can help to look at them separately.

example

Find the missing expressions from this addition grid:

+	$2x + 1$
$2x - 3$	$6x + 1$

The first column is straightforward, you just add the expressions:

$2x - 3 + 2x + 1$
$= 2x + 2x - 3 + 1$
$= 4x - 2$

For the second column, think of each part separately:

$2x - 3 + ? + ? = 6x + 1$
$2x + \text{④}x = 6x$ and $^-3 \text{④} = 1$
so $2x - 3 + \text{④}x \text{④} = 6x + 1$

The completed grid is:

+	$2x + 1$	$4x + 4$
$2x - 3$	$4x - 2$	$6x + 1$

Exercise A4.2

1 Use the inverse method to solve each of these equations:

a $5y - 4 = 26$	**b** $3y + 7 = 28$	**c** $4y - 11 = 29$	**d** $6y + 8 = 20$
e $2(y - 3) = 24$	**f** $7y + 3 = 38$	**g** $8y + 6 = 70$	**h** $3(y + 4) = 45$
i $2y - 18 = 50$	**j** $8(y + 5) = 56$	**k** $3y - 18 = 18$	**l** $7y + 8 = 50$
m $5(y + 9) = 65$	**n** $3(y - 5) = 9$	**o** $5y + 18 = 58$	**p** $12y - 8 = 28$

Use this code to change your answers to letters.

a	c	e	i	l	n	o	q	r	s	t	u
2	4	6	5	3	11	8	7	12	34	15	10

Read the letters as words – what does it say?

2 Copy and complete these addition grids:

a

+	8	x
7		$7 + x$
y		

b

+	$3x$	7
$2x$		
4		

c

+		5
	$8y$	
6	$2y + 6$	

d

+		
	$3x + 2y$	$2y + 5z$
6	$3x + 6$	

3 Copy and complete these addition grids:

a

+	$x + 3$	$x + 4$
$2x + 1$		$3x + 5$
$x - 2$		

b

+	$5x - 2$	$2x + 3$
$4x + 6$		
$7x - 8$		

c

+	$2x + 3$	
$3x + 8$		$5x + 12$
$4x - 5$		

d

+		
	$3x + 4$	$5x + 8$
	$5x + 6$	$7x + 10$

4 In an arithmagon, the expression in a square is the sum of the expressions in the two circles on either side of it.
Find the missing expressions.

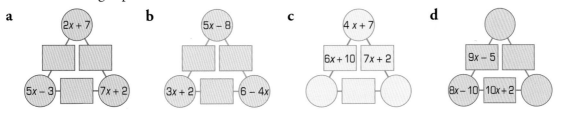

a (2x + 7), (5x – 3), (7x + 2)   **b** (5x – 8), (3x + 2), (6 – 4x)   **c** (4x + 7), 6x + 10, 7x + 2   **d** (9x – 5), (8x – 10), 10x + 2

Challenge

5 In these questions, use the information to make an equation.
Solve the equation to find the value of y.

a
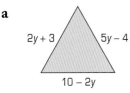
2y + 3, 5y – 4, 10 – 2y

Perimeter = 19 cm

b

2y, 74°, (y + 25)°

c
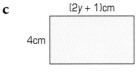
(2y + 1)cm, 4cm

Area = 28 cm^2

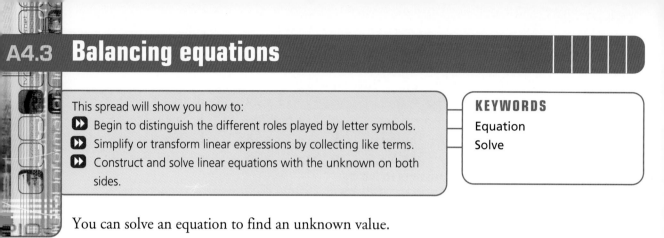

This spread will show you how to:
▶▶ Begin to distinguish the different roles played by letter symbols.
▶▶ Simplify or transform linear expressions by collecting like terms.
▶▶ Construct and solve linear equations with the unknown on both sides.

KEYWORDS
Equation
Solve

You can solve an equation to find an unknown value.

Solving equations is like a balancing act – you must always keep both sides of the equation equal or in balance.

The mass of each choc bar is unknown.

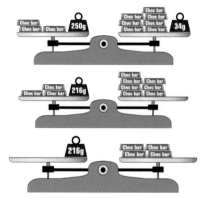

You can say the mass is m.

$3m + 250 = 7m + 34$

Take 34 g from both sides to keep the balance:

$3m + 216 = 7m$

Take $3m$ from both sides to keep the balance:

$216 = 4m$

If $4m = 216$ then $m = 216 \div 4 = 54$

The mass of a choc bar is 54 g.

example

Form two equations for this Hot Cross and solve them to find the value of y.

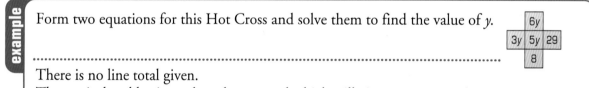

There is no line total given.
The vertical and horizontal totals are equal which will give you an equation.

Vertically: $6y + 5y + 8 = 11y + 8$
Horizontally: $3y + 5y + 29 = 8y + 29$

The totals are equal: $11y + 8 = 8y + 29$

Subtract 8 to keep the balance: $11y = 8y + 21$
Subtract $8y$ to keep the balance: $3y = 21$
                                        so $y = 21 \div 3 = 7$

Check the answer:

	42	
21	35	29	= 85
	8	

= 85

Exercise A4.3

1 Find the value of 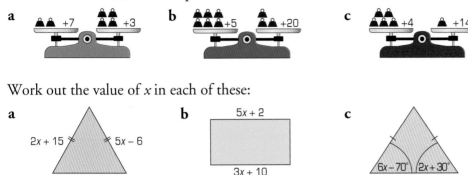 in each question:

a   ▲▲ +7   ▲▲ +3

b   ▲▲▲ +5   ▲▲ +20

c   ▲▲▲ +4   ▲ +14

2 Work out the value of x in each of these:

a

$2x + 15$ $5x - 6$

b

$5x + 2$

$3x + 10$

c

$6x - 70°$ $2x + 30°$

3 Solve each of these equations:

    a $3b + 7 = 8b - 18$ **b** $6c + 5 = 2c + 25$ **c** $8 + 3d = d + 26$

    d $4e - 7 = 3e - 3$ **e** $9f + 8 = 6f + 2$ **f** $15 + 3g = g + 20$

    g $6h - 18 = 2h + 14$ **h** $28 - 5i = 3i + 20$

4 In each Hot Cross the horizontal total is equal to the vertical total.
Form equations for each of these and solve them:

a
	2x	
6	5	4x
	3x	

b
	4x	
2x	5x	8
	2	

c
	8y	
5y	6	11
	5	

d
	2p	
6p	5p	6
	18	

5 Copy each of these algebra towers, add adjacent boxes together and
put the total in the box below.

a | $x + 7$ | $2x - 1$ | $3 - x$ |

b | $5x - 1$ | $5 - 3x$ | $4x + 2$ |

c | $6 - 2x$ | $3x - 4$ | $7 - x$ |

d | $2x + 4$ | $x + 5$ | $3x + 2$ |

6 In each of the towers in question **5** the tower total is equivalent to $2x + 32$.
Find the different values of x in each of the questions.

7 **Challenge**

    a In each of these towers you are given the tower total.
Use this to form an equation and solve it to find the value for x.

$2x + 7$		$x - 4$
	$3x - 7$	
	25	

$2x - 6$		$x - 4$
	$5x + 3$	
	$2x + 36$	

    b Make up your own tower challenge.

This spread will show you how to:
▶▶ Simplify or transform linear expressions by collecting like terms.
▶▶ Multiply a single term over a bracket.
▶▶ Construct and solve linear equations.

KEYWORDS
Multiply out   Solve
Algebraic expression

Algebra is as simple as arithmetic:

Compare 6×87 with $6(8t + 7)$:

$6 \times 87 = 6(80 + 7) = 6 \times 80 + 6 \times 7 = 480 + 42 = 522$
$6(8t + 7) = 6 \times 8t + 6 \times 7 = 48t + 42$

This is called expanding the bracket.

Remember:
$6 \times t = 6t$
$6 \times 2t = 12t$
$6 \times 3t = 18t$
and so on.

You can solve equations that include brackets if you expand them.

example

This flow diagram shows two different paths that lead to the same result.
Find the value of m.

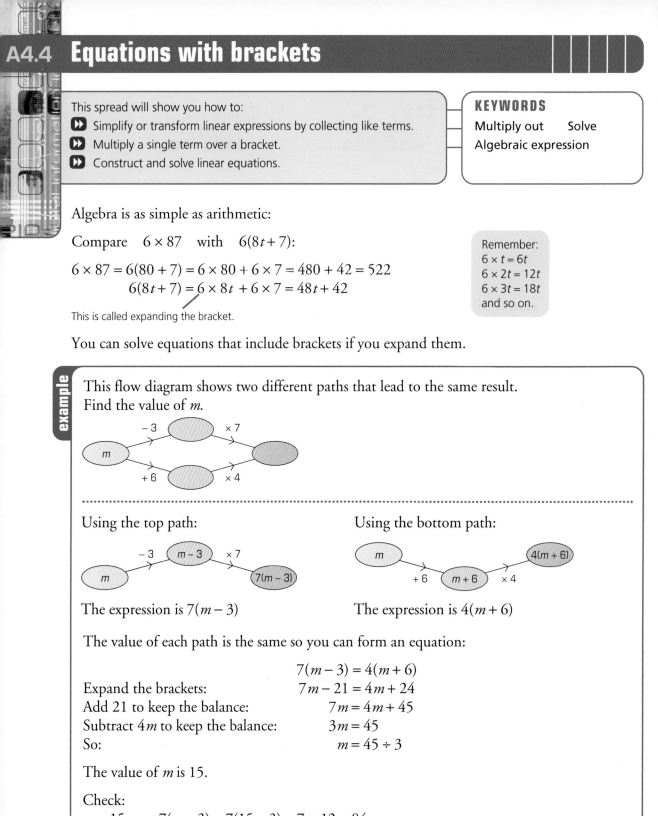

Using the top path:

The expression is $7(m - 3)$

Using the bottom path:

The expression is $4(m + 6)$

The value of each path is the same so you can form an equation:

$$7(m - 3) = 4(m + 6)$$

Expand the brackets: $7m - 21 = 4m + 24$
Add 21 to keep the balance: $7m = 4m + 45$
Subtract $4m$ to keep the balance: $3m = 45$
So: $m = 45 \div 3$

The value of m is 15.

Check:
$m = 15$ $7(m - 3) = 7(15 - 3) = 7 \times 12 = 84$
$4(m + 6) = 4(15 + 6) = 4 \times 21 = 84$

Exercise A4.4

1 Work out these multiplications. The first two questions are started for you.

a $178 \times 6 = 6 \times (100 + 70 + 8)$
$= 600 + 420 + 48$
$=$

b $39 \times 7 \quad = 7 \times (\qquad)$
$=$

c 143×9

d 391×4

e 684×5

2 Multiply out these brackets. The first one is started for you.

a $5(3a + 4) = 5 \times 3a + 5 \times 4 =$

b $6(4b - 3) =$

c $4(8 - 3c) =$

d $5(6d - 7) =$

e $7(11 - 6e) =$

f $4(8f + 3g) =$

g $3(2h - 8i) =$

h $4(5j - 6k + 4) =$

i $3(^-2 - 4l) =$

j $6(5 - 3m - 8n) =$

3 Match the algebraic expressions in the box with one or more of the expressions outside the box.

$5(4x - 10) \qquad 3(4x - 6) \qquad \frac{1}{2}(16x - 56)$

$7(6 + 9x)$

$3(15 + 18x)$

$2(6x - 9)$

$^-18 + 12x$

$10(2x - 5)$

$8x - 28$
$12x - 18$
$2(10x - 25)$
$42 + 63x$
$54x + 45$

$2(4x - 14)$

$6(2x - 3)$

$3(21x + 14)$

$9(6x + 5)$

$4(2x - 7)$

4 Solve these equations by first expanding the brackets:

a $3(x + 1) = 15$

b $4(2x + 1) = 20$

c $3(5 - x) = {}^-18$

d $5(1 - 2x) = 20$

5 Solve these two-way flow diagrams by forming equations.

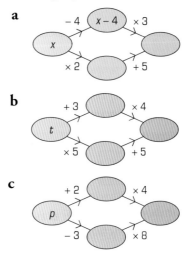

6 Solve these equations using the balance method. Expand the brackets first.

a $2(7x - 8) = 5x + 29$

b $3(8 - 5x) = 2x + 7$

c $3(5x - 4) = 4(2x + 4)$

d $2(7 - 5x) = 3(2x - 6)$

e $7x + 3(2x - 5) = 11$

f $3(2x + 7) = 7(x + 2)$

g $5(6 + 2x) = 2(x + 7)$

h $4(3x - 1) = 2(2x + 4)$

7 Solve these two-way flow diagrams by forming equations:

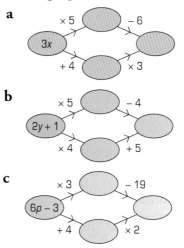

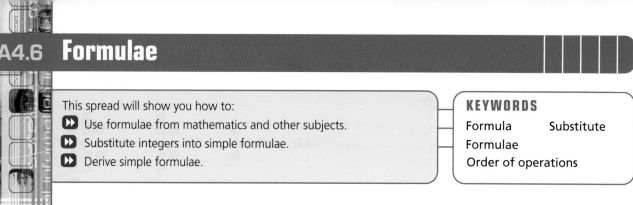

A4.6 Formulae

This spread will show you how to:
- ▶▶ Use formulae from mathematics and other subjects.
- ▶▶ Substitute integers into simple formulae.
- ▶▶ Derive simple formulae.

KEYWORDS

Formula   Substitute
Formulae
Order of operations

Brett is trying to work out $8 - 3 \times 2^2$:

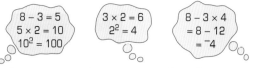

$8 - 3 = 5$
$5 \times 2 = 10$
$10^2 = 100$

$3 \times 2 = 6$
$2^2 = 4$

$8 - 3 \times 4$
$= 8 - 12$
$= {}^-4$

To find the correct answer, he must use the correct order of operations:

Brackets first:	$8 - 3 \times 2^2$	There are no brackets.
then		
Indices or powers	$8 - 3 \times 4$	$2^2 = 4$
then		
Divide and Multiply	$8 - 12$	$3 \times 4 = 12$
then		
Add and Subtract	${}^-4$	$8 - 12 = {}^-4$

These rules of arithmetic also apply to algebra.

If $a = 6$, $b = 5$ and $c = 3$, work out the value of:

a $a - bc$ **b** $(a + b)c$ **c** $\dfrac{4(3a - 5c)}{3}$ **d** $2ab - 3c^2$

....

Substitute the values into the expressions

a $a - bc$ **b** $(a + b)c$ **c** $\dfrac{4(3a - 5c)}{3}$ **d** $2ab - 3c^2$

$= 6 - 5 \times 3$ $= (6 + 5) \times 3$ $= \dfrac{4(3 \times 6 - 6 \times 3)}{3}$ $= 2 \times 6 \times 5 - 3 \times 3^2$

Multiply   Brackets   Brackets   Indices

$= 6 - 15$ $= 11 \times 3$ $= \dfrac{4(18 - 15)}{3}$ $= 2 \times 6 \times 5 - 3 \times 9$

$= {}^-9$ $= 33$ Multiply

                            $= \dfrac{4 \times 3}{3} = 4$ $= 60 - 27$

                                                         $= 33$

Exercise A4.6

1 On the school trip to Disneyland, Paris, Abigail has n euros and Bethan has m euros. Write expressions or equations for this information:

a Abigail and Bethan have 180 euros altogether.

b Abigail spends 45 euros on presents. How much has she left altogether?

c At the end of the day, Bethan has 30 euros left. How much has she spent?

2 In these questions, work out the value of each expression if

$$q = 7, \qquad r = 3, \qquad t = 4.$$

a $3r + 2t$ b $5q - 6r$ c $q^2 + r^2$

d $5(2r - t)$ e $t - 6r$ f $qt - \frac{1}{2}r$

g $\dfrac{3(5r - 3t)}{2}$ h $q(2r - 3)$ i $3qr - 2t^2$

3 In physics, you use the motion formula
$v = u + at$,
where v = final velocity (speed),
u = starting velocity, a = acceleration,
t = time.

a Find v if

   i $u = 3$, $a = 2$, $t = 5$

   ii $u = 7$, $a = 10$, $t = 5$

   iii $u = 10$, $a = 3$, $t = 4$

   iv $u = 10$, $a = {}^-2$, $t = 4$

b Find u if

   i $v = 20$, $a = 3$, $t = 5$

   ii $v = 36$, $a = 6$, $t = 4$

4 a Choose two numbers.

   ▶ Find their mean.

   ▶ Find the number halfway between the two numbers.

   What do you notice?

   Investigate for other pairs of numbers.

b Use your findings from part **a** to find a formula for the number, h, that is halfway between two numbers a and b.

5 A cube has six faces.

When two cubes are placed together you can count 10 faces (two faces are joined).

a How many faces can you count when 3 cubes and 4 cubes are placed together in a line?

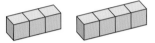

b Which of these formulae gives you the number of faces that can be seen when you place n cubes in a line?

   $2n + 4$ $6n - 2$ $4n + 2$ $8n - 2$

   Justify your answer.

c Use the formula to work out how many faces can be seen when you place 25 cubes together in a line.

6 In electronics, $V = IR$ where
V = voltage,
I = current,
R = resistance.

a Find the voltage when

   i $I = 15$, $R = 3$

   ii $I = 7$, $R = 1.5$

   iii $I = 2.5$, $R = 2$

b Find the resistance when

   i $V = 20$, $I = 4$

   ii $V = 24$, $I = 12$

   iii $V = 30$, $I = 2$

c Rewrite the formula with R as the subject.

   $R = _____$

You should know how to ...

1 Simplify or transform linear expressions by collecting like terms.

2 Multiply a single term over a bracket.

3 Substitute integers into simple formulae.

4 Represent problems in algebraic form, using correct notation.

5 Use logical argument to establish the truth of a statement.

Check out

1 Write different expressions for the total length of the lines in this diagram.

n

m

Simplify your expressions as far as possible. what do you notice?

2 Simplify these expressions
  a $4(x - 1)$
  b $2(3x + 5)$
  c $12 + 2(x - 3)$
  d $9 - (x + 5)$

3 The formula for the perimeter of a rectangle is:

$P = 2(l + w)$.

If P is 30 and w is 9, what is l?

4 Use algebra to write this statement correctly:
'I think of a number, add 6 then multiply by 2. The answer is 18.'
Now solve the equation to find the value of the number.

5 Prove that the sum of any three consecutive numbers is always divisible by 3.

Hint: Call the first number n, then the next number is $n + 1$.

This unit will show you how to:

▶▶ Discuss a problem that can be addressed by statistical methods and identify related questions to explore.

▶▶ Decide which data to collect to answer a question, and the degree of accuracy needed.

▶▶ Identify possible sources of data.

▶▶ Plan how to collect data, including sample size.

▶▶ Collect data using a suitable method.

▶▶ Calculate statistics, including with a calculator.

▶▶ Recognise when it is appropriate to use the range, mean, median and mode.

▶▶ Construct a stem-and-leaf diagram.

▶▶ Communicate the results of a statistical enquiry and the methods used.

▶▶ Construct charts on paper. Identify which are most useful in the context of the problem.

▶▶ Interpret tables, graphs and diagrams for discrete data, and draw inferences that relate to the problem being discussed.

▶▶ Relate summarised data to the questions being explored.

▶▶ Identify information to solve a problem.

▶▶ Solve more complex problems by breaking them into smaller steps or tasks.

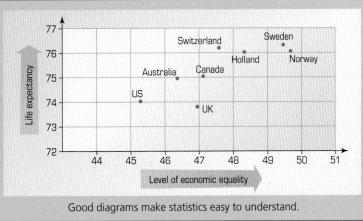

Good diagrams make statistics easy to understand.

Before you start

You should know how to ...

1 Construct a bar chart for discrete grouped data.

2 Find the mean, median, mode and range.

▶ The mean = $\dfrac{\text{the sum of the values}}{\text{the number of values}}$

▶ The mode is the most common value.

▶ The median is the middle value when the data is arranged in size order.

▶ The range = highest value – lowest value

Check in

1 Draw a bar chart for these results of a survey into films seen.

Films seen	0–4	5–9	10–14	15 and above
Frequency	2	7	13	3

2 24 people were asked how many items of mail they received one morning.
The results are:

7, 4, 3, 2, 6, 2, 3, 3, 6, 4, 5, 2,
5, 5, 4, 7, 5, 2, 3, 4, 1, 0, 5, 6

Find the mean, median, mode and range.

This spread will help you to:
▶▶ Discuss a problem that can be addressed by statistical methods and identify related questions to explore.
▶▶ Decide which data to collect to answer a question.
▶▶ Identify possible sources of data.
▶▶ Plan how to collect the data.
▶▶ Collect data using a suitable method.

KEYWORDS
Sample   Plan
Primary data
Secondary data

Kate and Ali are at a football match.
They want to know the best time to go to the food bar to:
▶ avoid the half-time queues but
▶ ensure they don't miss a goal.

They need to collect data to answer the question.
They have to consider whether to use:
▶ primary data – how will they collect it? or
▶ secondary data – what is the best source?

They use the times of goals in the weekend football results.

This is secondary data.

It's a good source as it's:
▶ easily available and
▶ clear and easy to use.
The sample size is large enough to draw conclusions.

They group the times into 15-minute intervals so the data is easy to analyse.

Time (min)	Frequency
1–15	8
16–30	35
31–45	22
46–60	20
61–75	42
76–90	30

They draw a bar chart to help them see the patterns:

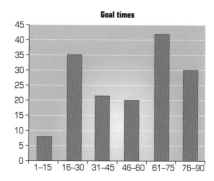

Goal times

The data suggests:
▶ The 'middle third' of each half of the game is the peak time for goals to be scored.
▶ The least likely time for a goal to be scored is right at the start of the match.
▶ More goals were scored in the second half than in the first half.

Next week's game is in the Cup – Kate and Ali want to know if the pattern will be any different. They will collect the data and analyse it again.

Exercise D2.1

Each set of data is accompanied by a hypothesis.
a Write a paragraph about each set.
b Decide whether the data given would let you test the hypothesis.
c If you think that you would need other data, explain what this
   would be, and where you could get it.

1

Hypothesis: Men from Hightown are taller than men from Deepdale.	
Height (in cm) of 10 men from Hightown	Height (in cm) of 10 men from Deepdale
168, 178, 189, 167, 175, 176, 188, 169, 175, 188	167, 164, 188, 174, 164, 161, 173, 174, 159, 172

2

Hypothesis: SupaBrite light bulbs last longer than EverGlow light bulbs.	
100 bulbs of each type were run continuously in a laboratory. The machinery used recorded how many hours each light bulb ran before it failed. A summary of the results is shown below.	
SupaBrite light bulbs Mean = 1304 hours Range = 409 hours	EverGlow light bulbs Mean = 1116 hours Range = 229 hours

3

Hypothesis: The month of June was sunnier in Sinkwell Bay than it was in Fairdrift Cove.

The table shows the June sunshine record for each resort.

Hours of sunshine	Number of days – Sinkwell Bay	Number of days – Fairdrift Cove
6–7	1	5
5–6	3	7
4–5	7	3
3–4	9	2
2–3	5	1
1–2	3	4
0–1	2	8

0–1 hours means up to but not including 1 hour.

4

Hypothesis: School students spend more time playing sports than adults do.

100 adults and 100 school students were asked how long they spent playing sports in the last week. Their answers are shown in the chart.

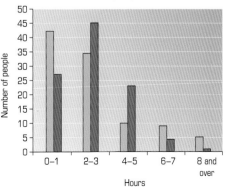

Time spent playing sports

Key:
☐ Adults
▨ Students

This spread will help you to:
▶▶ Construct pie charts for categorical data.
▶▶ Construct bar charts and frequency diagrams for discrete data.
▶▶ Construct simple scatter graphs.
▶▶ Identify which charts are most suitable.

KEYWORDS
Pie chart   Scatter graph
Discrete

▶ A pie chart uses a circle to display discrete data.
  It compares the size of a category with the whole.

A bar chart compares the sizes of categories with each other.

For any category or sector in a pie chart you:

▶ work out the fraction of the whole and then
▶ find this fraction of 360°.

This gives the angle you use.

$$\text{Angle for category} = \frac{\text{Value of category}}{\text{Total of all categories}} \times 360°$$

To draw the pie chart for any data, you work out the angles needed:

Pizza	Number ordered	Sector angle
Hot 'n' Spicy	5	$\frac{5}{23} \times 360° = 78°$
Super Combo	10	$\frac{10}{23} \times 360° = 157°$
Ocean Treat	4	$\frac{4}{23} \times 360° = 63°$
Big Cheese	3	$\frac{3}{23} \times 360° = 47°$
Hawaiian Holiday	1	$\frac{1}{23} \times 360° = 16°$

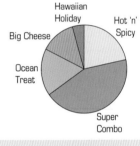

The angles are rounded so they don't add to 360°.

▶ A scatter graph is useful for comparing two sets of data with each other.

The table shows the heights and weights of Gemma's classmates.

Name	Height (cm)	Weight (kg)
Mark	154	61
Karl	168	72
Phil	148	56
Kam	153	52
Derek	175	80
Nina	160	66
Gill	157	58
Pete	167	84

The points lie fairly close together so the scatter is not very wide.

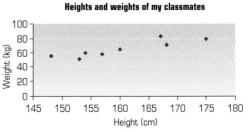

The graph makes the data easier to analyse.

Exercise D2.2

1 100 clients were asked to rate a company's service. The table shows their responses.

Rating	Excellent	Good	Fair	Poor
Number	38	45	12	5
Pie chart angle				

Copy and complete the table, and show the data as a pie chart.

2 Draw scatter diagrams for these sets of data.
Comment on the spread of the data.

  a Two wine tasters were asked to taste 10 different wines.
     They gave each one a mark out of 10.

Wine	A	B	C	D	E	F	G	H	I	J
Taster 1	5	7	1	9	3	3	5	7	8	1
Taster 2	4	6	3	9	4	2	5	8	7	1

  b The owner of a café kept a record of the number of ice-creams
     and hot dogs sold each day, for 10 days.

Day	1	2	3	4	5	6	7	8	9	10
Ice-creams	42	61	51	43	32	27	11	37	64	82
Hot dogs	53	32	40	47	61	75	84	62	31	9

3 Jenny collected this data for her project.
She wants to illustrate the data using appropriate diagrams.
Recommend the best diagram for her to use in each case.
Justify your recommendations.

Attendance at the local cinema:

Day	Mon	Tue	Wed	Thu	Fri	Sat	Sun
Attendance	110	126	98	48	145	145	85

Age ratings of film released this month:

Age rating	U	12	15	18
Number of films	10	4	7	9

Two people's order of preference for 5 films:
    5 = the best, 1 = the worst

Film	A	B	C	D	E
Alan	5	2	4	1	3
Brenda	3	1	4	2	5

4 Colin started to work out the angles needed for a pie chart to display the data in the table.
Unfortunately, he spilt some coffee on the paper.

Species	Sparrow	Blackbird	Pigeon	Blue tit	Wagtail
Number	12	6	8	17	
Angle	80°				

Copy and complete the table, and draw the pie chart for the data.

This spread will help you to:
▶▶ Interpret tables, graphs and diagrams for discrete data, and draw inferences that relate to the problem being discussed.
▶▶ Relate summarised data to the questions being explored.

KEYWORDS
Interpret
Population pyramid
Pie chart

You need to decide which inferences you can make, based on the diagram – and which you can't!

An **inference** is a conclusion based on reasoning.

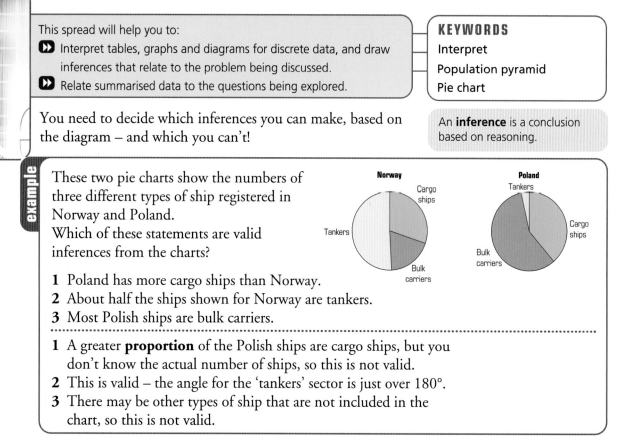

These two pie charts show the numbers of three different types of ship registered in Norway and Poland.
Which of these statements are valid inferences from the charts?

1 Poland has more cargo ships than Norway.
2 About half the ships shown for Norway are tankers.
3 Most Polish ships are bulk carriers.

....

1 A greater **proportion** of the Polish ships are cargo ships, but you don't know the actual number of ships, so this is not valid.
2 This is valid – the angle for the 'tankers' sector is just over 180°.
3 There may be other types of ship that are not included in the chart, so this is not valid.

A **population pyramid** shows the differences between males and females in a population.

The diagrams are population pyramids for Jersey and Ireland.

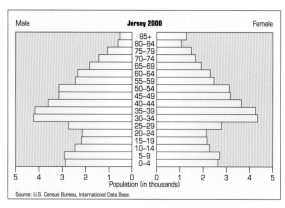

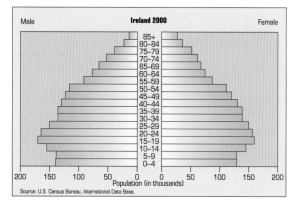

The main features are:
▶ The population of Ireland is much larger than that of Jersey.
▶ There is a greater proportion of young people in Ireland.
▶ In both places, women outnumber men significantly after about 70 years.

Exercise D2.3

For each question:

a Write a paragraph explaining the main features of the diagrams.
b Explain how you think the data was collected.
c Decribe any ways in which the diagram could be wrongly interpreted.

1 Urban and rural populations.

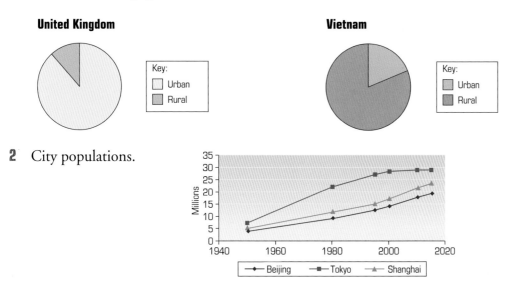

2 City populations.

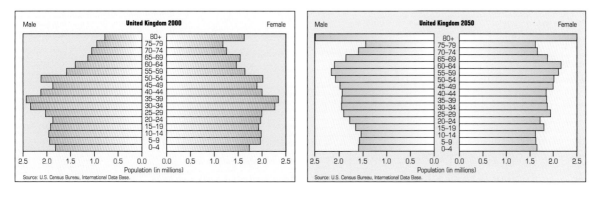

3 United Kingdom population pyramids, actual and predicted populations.

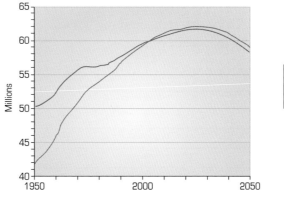

4 Actual and predicted populations – UK and France

This spread will help you to:
▸▸ Calculate statistics.
▸▸ Construct and use a stem-and-leaf diagram.

KEYWORDS
Stem-and-leaf diagram
Mode   Median
Modal class   Range

Stem-and-leaf diagrams are a good way of organising numerical data.
They are similar to grouped frequency diagrams but keep the raw data.

The table shows the heights, in centimetres, of 25 Year 8 students:

143	170	135	150	158
169	154	150	157	167
154	155	158	171	169
172	161	142	157	162
154	138	145	153	158

▸ Order the data and then choose an appropriate stem.
  In this case, using multiples of 10 is appropriate: 130, 140, 150, 160, 170.
▸ Order the data on the leaves:

130	5	8									
140	2	3	5								
150	0	0	3	4	4	4	5	7	7	8	8	8
160	1	2	7	9	9						
170	0	1	2								

The bottom row shows students with heights of 170 cm, 171 cm and 172 cm.

The diagram shows the shape of the distribution.

It is easy to read off the mode, median and range.

▸ There are two **modes** – 154 cm and 158 cm. They have the most entries.
▸ The modal class is 150 cm to 160 cm. It has the longest leaf.
▸ The **median** is the 13th value, which is 157 cm – count from the top.
▸ The **range** is 172 cm−135 cm = 37 cm

You can choose scales for the 'stem' and the 'leaves'.
You should use a key to explain your scale.

This diagram shows the times taken to run a 200 m race by a squad of 24 athletes.

22	2	5	6	8	8	9
23	1	3	3	3	6	7	8
24	2	4	7	9	9	
25	7	8	8	9		
26	4	5				

Key: | 23 | 6 |
means 23.6 seconds

The median is 23.75s

Exercise D2.4

1 The table gives the heights of 21 people in centimetres.
The heights have been put into order.

101	113	119	121	124	124	126
130	131	134	137	138	140	145
145	147	150	155	156	162	171

Use the data to copy and complete the stem-and-leaf diagram.
The first three values have already been entered.

100	1	
110	3	9
120		
130		
140		
150		
160		
170		

Key:

100	1

represents 101 cm

2 This table shows the weights of 27 people in kilograms.
The weights have **not** been put into order.

49	81	53	51	64	58	41	42	63
55	60	74	63	57	44	36	52	68
72	45	37	57	34	59	40	65	52

  a Draw a stem-and-leaf diagram for this set of data.
    Use multiples of 10 as the 'stem' and units as the 'leaves'.
    Include a key to explain how the diagram works.
  b Use the diagram to find the median.

3 The ages of 20 passengers on a bus are:

36, 12, 12, 23, 24, 1, 25, 1, 46, 17, 37, 12, 31, 4, 8, 51, 13, 14, 25, 28.

  a Draw a stem-and-leaf diagram for the ages.
  b Find the median age of the passengers.
  c Find the range of the ages of the passengers.

4 Here are three sets of data.

The heights of 26 trees (in metres, to 1 dp):

4.6, 3.2, 5.4, 2.4, 3.5, 7.7, 3.8, 3.7, 1.3, 7.4, 4.9, 3.3, 4.3, 2.8, 6.5, 2.4, 5.0, 3.9, 3.2, 1.7, 8.1, 2.3, 5.7, 4.1, 7.6, 6.1

The weights of 35 puppies (in grams).

609, 595, 618, 602, 600, 596, 582, 571, 624, 581, 554, 570, 657, 624, 617, 561, 547, 642, 584, 566, 627, 627, 616, 571, 664, 626, 638, 599, 632, 624, 603, 598, 550, 577, 568

The temperatures (in °C) of metal samples in a laboratory.

395, 366, 368, 361, 368, 388, 366, 361, 356, 369, 384, 371, 337, 370, 363, 372, 386, 366, 360, 360, 372, 374, 361

For each set of data:
  a Draw a stem-and-leaf diagram.
  b Find the median and range.
  c Use your stem-and-leaf diagram to identify the modal class.

This spread will help you to:
▶▶ Calculate statistics, including with a calculator.
▶▶ Recognise when it is appropriate to use the range, mean, median and mode.

KEYWORDS
Range Mean Distribution
Median Mode Average
Stem-and-leaf diagram

The overall shape of a chart or diagram often gives a good picture of a set of data.

The mean, median, mode and range are examples of **statistics**.

You can calculate statistics to summarise the picture.

The marks scored by two classes in a maths test are shown in this **back-to-back stem-and-leaf diagram**.

		Class A						Class B					
			6	4	1	0	40	3	8				
8	7	7	7	5	2	2	50	2	4	5	7		
9	8	8	4	4	3	1	60	0	2	4	4	6	7
		8	8	8	3	2	70	4	4	4	6	7	8	8
					6	1	80	3	5	5	7	9	
						1	90	2	4	7			

You can compare the two classes' performance in various ways:

Key:
means 43

▶ Look at the overall shape of the **distribution**:
The stem-and-leaf diagram shows that the marks in Class B were generally higher than those in Class A.
▶ Use an **average**:

	Mean	Median	Mode
Class A	63.6	63.5	78
Class B	71.7	74	74

You should know how to calculate each of these averages.

▶ Find the **range** of marks:
For Class A this is $91 - 40 = 51$ marks
The range for Class B is $97 - 43 = 54$ marks

You need to consider which average gives the fairest picture.

It would be better to find the modal class.

▶ The **mode** for Class A is higher than the mode for Class B but this does not reflect the general picture.
▶ The **mean** uses all the data, and so if the student with the highest score moves class, the mean score will change.
▶ The **median** will not be affected so much by a change in the data.
▶ If the maximum or minimum value changes so will the **range**, even if all the other data values stay the same.

Exercise D2.5

1 The weights (in kg) of a set of nine items of airline luggage are:

13, 21, 7, 29, 25, 16, 17, 9, 20.

   a Find the **mean** weight of the items.
   b Find the **median** weight
   c Explain why there is no **mode**.
   d Find the **range** of the weights.

2 The weight of the heaviest item of luggage in question **1** is reduced to 25 kg.
   a Explain the effect that this change will have on the three different averages.
   b Explain the effect that the change will have on the range of the weights.

3 The midday temperatures (in °C) at 10 different weather stations are:

Station	A	B	C	D	E	F	G	H	I	J
Temp.	19	23	15	45	26	11	15	21	24	26

   a Find the mean, median, mode and range for this data.
   b The temperature reported by station D was found to be incorrect – it should have been 25 °C, not 45 °C. Recalculate the mean, median, mode and range for the corrected data.

4 The heights (in cm) of the children in two groups are shown in the table.

Group A	Group B
147, 162, 135, 143, 156, 154, 138, 143, 161, 159, 142	152, 134, 145, 134, 146, 157, 133, 161, 141, 121, 137, 128

   a Draw a back-to-back stem-and-leaf diagram for the data.
   b Find the range of each set of heights.
   c Find the median height for each set.
   d Find the mean of each set of heights.
   e Use your diagram to find the modal class for the heights of each set.

5 In question **4**, the shortest person in set B is moved into set A. Explain the effect this will have on each of the statistics which you calculated in parts **b**, **c**, **d** and **e**.

6 A set of 7 number cards is marked as shown.

   a Find the mean, median, mode and range of the numbers on the cards.

One of the cards is removed, to leave a set of 6. Explain which card you should remove to keep these statistics unchanged (there may be more than one possibility):
   b The **range**
   c The **mean**
   d The **median**
   e The **mode**
   f All of **b–e**.

7 The data in question **6** had an unusual property: you could remove one of the items, and the new set would have exactly the same mean, median, mode and range.

Construct data sets with these properties:
  ▶ **Set A**: Set A has a mode. Whichever item you remove, the mode remains the same.
  ▶ **Set B**: You can remove one of the items of Set B, so that the mean of the new set is double the mean of the original set.
  ▶ **Set C**: It doesn't matter which item you remove from Set C – the mean, median, mode and range will **all** change.

Make up some more puzzles of this type, and investigate whether it is possible to solve them.

This spread will help you to:
- ▸ Discuss a problem that can be addressed by statistical methods and identify related questions to explore.
- ▸ Communicate the results of a statistical enquiry and the methods used, justifying the choice of what is presented.

KEYWORDS

Hypothesis   Justify
Conclusion
Continuous

Once you have undertaken a statistical enquiry, you need to report your results.

> Hypothesis: Overall, Year 8 boys are taller than Year 8 girls.

Data collection

We wanted a large and representative set of data so we downloaded this data from the Census at School website.

We explain **what** data was collected, **why**, and **how**.

Height (cm)	Boys	Girls
91–100	1	2
101–110	3	6
111–120	11	21
121–130	43	31
131–140	185	173
141–150	1126	1101

Height (cm)	Boys	Girls
151–160	1665	2136
161–170	668	819
171–180	102	74
181–190	11	10
191–200	2	8
Total	3817	4381

> Analysis of data
> The modal class for both boys and girls was 151–160 cm.
> The median value is in the 151–160 cm interval in both cases.
> The range was the same in both cases.

Graphs to represent data

The actual heights are continuous so we will use a frequency diagram. The graphs show there are more girls taller than 150 cm than boys.

Conclusions

There is no evidence to suggest that Year 8 boys are taller than Year 8 girls. If anything, there might be some evidence to support the opposite.

Areas for further investigation

We could investigate the situation for other age groups.

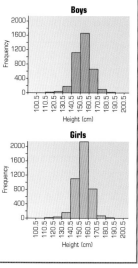

Exercise D2.6

Here are brief details of two statistical projects.

1 Books

As part of a project for World Book Week, a class decides to compare the time Year 8 students spend reading and playing computer games.

Hypothesis: 'We are going to find out about the time people spend reading and playing computer games.'

Data collection: A questionnaire is given to each member of the class; they record how many hours are spent on each activity in a month.

Data:

Hours spent reading books:
26, 8, 19, 18, 25, 35, 17, 9, 32, 21, 14, 26, 11, 15, 23, 34, 28, 11, 18, 22, 13

Hours spent playing computer games:
27, 36, 15, 0, 3, 24, 25, 0, 36, 0, 16, 25, 0, 29, 35, 13, 0, 15, 37, 28, 17

2 Fertilizer

A television gardening programme is testing a tomato plant fertilizer to decide whether to recommend it to the viewers.

Hypothesis: 'GroFast fertilizer makes tomato plants produce larger tomatoes than untreated plants.'

Data collection: 10 tomato plants were grown with GroFast fertilizer, and 10 without. The tomatoes from each plant were picked and weighed.

Data:

Weight of tomatoes (kg) – no fertilizer:
5.3, 4.1, 4.3, 2.4, 4, 5, 3.8, 5.3, 3.6, 5.1

Weight of tomatoes (kg) – with GroFast fertilizer:
6.5, 5.7, 6.5, 3.6, 4.5, 4.8, 4.6, 4.9, 5.7, 4.8

For each of these projects write a paragraph about each these points:

a The Hypothesis

How clear is the hypothesis?

Does it clearly set up an idea that the project will set out to test?

Can you think of a better hypothesis for the project?

b Data collection

Does the project have a sensible approach to gathering data?

Consider the sample size – how much data is needed to reach reliable conclusions?

Think about any precautions that will be needed in collecting the data.

Are there any problems that might occur – and how can they be avoided?

Now complete these stages for each of the projects:

c Analysing and presenting data

Draw appropriate graphs and calculate any statistics you need.

d Conclusions

Decide whether each hypothesis should be accepted or rejected.

Give reasons for your decision.

e Areas for further investigation

Write about any appropriate developments.

This spread will show you how to:

▶▶ Consolidate and extend mental methods of calculation for addition and subtraction.

▶▶ Understand addition and subtraction of fractions and integers.

▶▶ Consolidate standard column procedures for addition and subtraction of integers and decimals.

KEYWORDS

Fraction	Positive
Decimal	Negative
Equivalent	Commutative

The laws of arithmetic for addition and subtraction are the same for all numbers: positive, negative, fractions and decimals.

▶ Addition is the inverse of subtraction:

$65 - 27 = 38$	so	$38 + 27 = 65$
$0.75 - 0.2 = 0.55$	so	$0.55 + 0.2 = 0.75$
$\frac{3}{4} - \frac{1}{5} = \frac{11}{20}$	so	$\frac{11}{20} + \frac{1}{5} = \frac{3}{4}$
$65 - {}^-27 = 92$	so	$92 + {}^-27 = 65$

▶ Addition is commutative, but subtraction is not:

$38 + 27 = 27 + 38$	but	$65 - 27 \neq 27 - 65$
$0.55 + 0.2 = 0.2 + 0.55$	but	$0.75 - 0.2 \neq 0.2 - 0.75$
$\frac{11}{20} + \frac{1}{5} = \frac{1}{5} + \frac{11}{20}$	but	$\frac{3}{4} - \frac{1}{5} \neq \frac{1}{5} - \frac{3}{4}$
$92 + {}^-27 = {}^-27 + 92$	but	$65 - {}^-27 \neq {}^-27 - 65$

To add or subtract decimals, you can add zeros after the decimal point.

$0.87 + 160.9 + 2365 =$

$$
\begin{array}{r}
0.87 \\
160.90 \\
2365.00 \\
\hline
2526.77
\end{array}
$$

The zeros add no value but help you set out the calculation.

Estimate:
$0.87 + 160.9 + 2365$
$\approx 160 + 2360$
$= 2520)$

To add or subtract fractions, change them to equivalent fractions with a common denominator, or to their equivalent decimals.

example

Find the sum of $\frac{2}{5} - \frac{3}{8} + \frac{11}{20}$

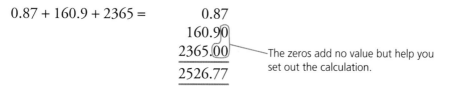

Using equivalent fractions:

$\frac{2}{5} = \frac{16}{40}$

$\frac{3}{8} = \frac{15}{40}$

$\frac{11}{20} = \frac{22}{40}$

$$\frac{2}{5} - \frac{3}{8} + \frac{11}{20} = \frac{16 - 15 + 22}{40} = \frac{23}{40}$$

Using decimals:

$\frac{2}{5} = 2 \div 5 = 0.4$

$\frac{3}{8} = 3 \div 8 = 0.375$

$\frac{11}{20} = 11 \div 20 = 0.55$

$= 0.4 - 0.375 + 0.55 = 0.575$

Exercise N4.1

1 Calculate using a mental or written method:

  a $4.8 + 2.4$ **b** $58 + {}^{-}29$

  c $26.3 - 5.27$ **d** $164.7 - 11 + 17.2$

2 Calculate these using a mental or written method:

  a $13.08 + 2 + 0.05$

  b $17.3 - 1.47 + {}^{-}0.7$

  c $43.6 - {}^{-}185.7 + 63.53$

  d $4.09 + 2.7 + {}^{-}6 + 0.89$

  e $178 - 2.078 - 3.9$

  f $0.0235 + 140.002 + 3248$

  g $243.05 + 0.049 - 16.74$

  h $4.082 \text{ km} - {}^{-}0.068 \text{ km}$

3 **a** Copy the diagram and write down fraction facts you can say because $\frac{2}{5} + \frac{1}{2} = \frac{9}{10}$:

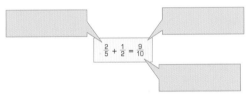

  b Write some more facts based upon $\frac{2}{5} + \frac{1}{2} = \frac{9}{10}$ using negative numbers and decimals.

4 **Puzzle**

In a pyramid, the brick that sits directly above two bricks is the sum of those two bricks:

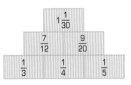

Copy and complete these pyramids:

  a   **b**

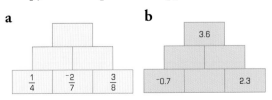

5 **a** Joey, Ross and Chandler buy two identical pizzas to share.
Ross has $\frac{1}{4}$ of the first pizza and $\frac{3}{8}$ of the second pizza;
Chandler has $\frac{1}{3}$ of the first pizza and $\frac{13}{40}$ of the second pizza;
Joey has $\frac{5}{12}$ of the first pizza and $\frac{3}{10}$ of the second pizza.
Who had the most pizza?
Explain and justify your answer.

  b The difference between two numbers is $\frac{9}{20}$.
The sum of the two numbers is $1\frac{1}{5}$.
What are the two numbers?
Explain your method.

6 **Puzzle**

Here are six numbers.

15.3 ${}^{-}12.07$ ${}^{-}32.4$

0.4 ${}^{-}0.67$ 19.07

> To make 2.83 with three numbers, use:
> $15.3 + {}^{-}12.07 - 0.4$

You can add or subtract the numbers to make a target number.

  a Find the two numbers that can be used to make 31.73

  b Find the three numbers that can be used to make 31.81

  c Find the four numbers that can be used to make ${}^{-}33.3$.

7 **Puzzle**

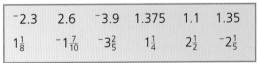

Use the twelve numbers in the box to complete these number sentences:

  a ☐ + ☐ = ☐

  b ☐ − ☐ = ☐

  c ☐ + ☐ = ☐

  d ☐   ☐ = ☐

> Use each number only once.

This spread will show you how to:

▶▶ Consolidate and extend mental methods of multiplication and division.

▶▶ Understand multiplication and division of integers.

▶▶ Use the laws of arithmetic and inverse operations.

KEYWORDS

Division   Negative

Positive   Commutative

Multiplication

Inverse

The laws of arithmetic for multiplication and division are the same for all numbers: positive, negative, fractions and decimals.

▶ **Division is the inverse of multiplication.**

$12 \times 9 = 108$	so	$108 \div 9 = 12$
$12 \times \frac{2}{5} = 4\frac{4}{5}$	so	$4\frac{4}{5} \div \frac{2}{5} = 12$
$12 \times 0.4 = 4.8$	so	$4.8 \div 0.4 = 12$
$12 \times {}^-9 = {}^-108$	so	${}^-108 \div {}^-9 = 12$

▶ **Multiplication is commutative, division is not.**

$12 \times 9 = 9 \times 12$	but	$108 \div 9 \neq 9 \div 108$
$12 \times \frac{2}{5} = \frac{2}{5} \times 12$	but	$4\frac{4}{5} \div \frac{2}{5} \neq \frac{2}{5} \div 4\frac{4}{5}$
$12 \times 0.4 = 0.4 \times 12$	but	$4.8 \div 0.4 \neq 0.4 \div 4.8$
$12 \times {}^-9 = {}^-9 \times 12$	but	${}^-108 \div {}^-9 \neq {}^-9 \div {}^-108$

You can multiply any fraction by an integer.

For example:
$$\frac{4}{5} \times {}^-17 = (\frac{1}{5} \times 4) \times {}^-17$$
$$= \frac{4 \times {}^-17}{5}$$
$$= \frac{{}^-68}{5}$$
$$= {}^-13\frac{3}{5}$$

Multiplying by $\frac{1}{5}$ = Dividing by 5

You can multiply and divide by multiples of 0.1 and 0.01 using factors.

example

Calculate:

a ${}^-17 \times 0.8$ **b** ${}^-3.4 \div 0.1$ **c** $12 \div 0.4$ **d** ${}^-24 \div 0.06$

....

a ${}^-17 \times 0.8$
$= {}^-17 \times (8 \times 0.1)$
$= {}^-136 \times 0.1$
$= {}^-13.6$

b ${}^-3.4 \div 0.1$
$= {}^-3.4 \div \frac{1}{10}$
$= {}^-3.4 \times 10$
$= {}^-34$

c $12 \div 4 = 3$
$3 \div 0.1 = 30$
$12 \div 0.4 = 30$

d ${}^-24 \div 6 = {}^-4$
${}^-4 \div 0.01 = {}^-400$
${}^-24 \div 0.06 = {}^-400$

Exercise N4.2

1 Calculate mentally:

   a 17×11 **b** 0.9×8

   c $^-4 \times 12$ **d** $108 \div 9$

   e 3×1.4 **f** 26×19

   g $235 \div 0.1$ **h** 0.3×20

   i 37×0.01 **j** $192 \div 6$

   k $\frac{1}{7} \times 14$ **l** $40 \times \frac{1}{8}$

2 **a** Parking costs 20p for every 15-minute period. How much will it cost to park a car from 8.45 am until 3.30 pm?

   b On Thursday the temperature dropped 12.6 degrees. On Friday the temperature only dropped by $\frac{1}{3}$ of the amount it dropped on Thursday. How much did the temperature fall on Friday?

   c The entrance fee for a museum is £2.20 per person. How much is the total cost for a party of 14 people to visit the museum?

3 In each multiplication trail you must choose numbers, one from each row, which multiply together to give the target number.

For example:

0.2	0.4	0.3
13	15	12
14	16	17

	48	

> $0.2 \times 15 = 3$
> $3 \times 16 = 48$
> So $0.2 \times 15 \times 16 = 48$

a

0.3	0.4	0.6
17	13	15
19	21	11

	74.1	

b

0.4	0.8	0.6
21	11	31
9	19	29

	255.2	

4 These are the capacities of some objects you might find in a kitchen:

Mustard jar	*0.05 litre*
Cup	*0.25 litre*
Milk jug	*0.5 litre*
Fruit juice carton	*1 litre*
Saucepan	*2.4 litres*
Bucket	*6 litres*

Write down five statements about the relative capacities of these objects.

For example:

'It would take 12 milk jugs to fill a bucket'

'The mustard jar is $\frac{1}{5}$th of the capacity of a cup'.

5 You may need to make some jottings to calculate:

   a $^-15 \times 0.6$ **b** $^-28 \div 0.7$

   c $\frac{3}{8} \times {}^-24$ **d** $^-4 \div {}^-0.02$

   e $\frac{2}{5} \times 28$ m   **f** $^-1.6 \div 0.2$

   g $^-18 \times 3.1$ **h** 6.8×2.5

6 **a** Find the value of each of these expressions if $x = 0.4$, $y = {}^-2.5$ and $t = 12$.

     i $2x$ **ii** $4y$ **iii** $2t + 6$ **iv** $2x - 4y$

     v $t^2 - 20y$ **vi** $\dfrac{3t}{x}$ **vii** $\dfrac{5x}{y}$

   b Make up five expressions of your own with a value of $^-5$ using x, y and t. For example $2y = {}^-5$.

7 **Puzzle**

In negative countdown you must use all the numbers to make the given target number. You may add, subtract, multiply or divide the numbers.

   a Target = 300 $^-42$ 21   0.7   3 $^-12$

   b Target = 32 $\frac{3}{5}$ $^-10$ 0.4 $^-3.5$ $^-8$

This spread will show you how to:

▶▶ Use the order of operations, including brackets, with more complex calculations.

▶▶ Carry out more difficult calculations effectively and efficiently using the function keys of a calculator.

A scientific calculator works out calculations using the mathematical order of operations. You need to know and use this order in complex calculations.

example

Calculate $\dfrac{(1.8 + 1.2)^2}{\sqrt{(8 + 4 \times 7)}} + 3 \times 0.5$

You should be able to do this in your head using jottings!

Order of operations:

$\dfrac{(1.8 + 1.2)^2}{\sqrt{(8 + 4 \times 7)}} + 3 \times 0.5$

Brackets:
Work out the contents of the brackets first.

↓

Powers or Indices:
Work out powers or roots next.

$= \dfrac{3^2}{\sqrt{36}} + 3 \times 0.5$

↓

Multiplication and Division
Work out any multiplications and divisions.

$= \dfrac{9}{6} + 3 \times 0.5$

↓   ↓

Addition and Subtraction
Finally, work out additions and subtractions.

$= 1.5 + 1.5$

$= 3$

On a calculator you can input the calculation all in one go but it is easy to make a mistake:

$(1.8 + 1.2)\,\boxed{x^2} \div \boxed{\sqrt{}}$

$(8 + 4 \times 7) + 3 \times 0.5 =$

You use $\boxed{x^2}$ to square a number.

press 3 $\boxed{x^2}$ = 9

Careful! On some calculators you use the $\boxed{\sqrt{}}$ key after the number:

Input $(8 + 4 \times 7)\,\boxed{\sqrt{}} =$

Check you know which way your calculator works.

To change $\frac{9}{6}$ to a decimal, input

9 $\boxed{x/y}$ 6 = the output should show 1⌊1⌊2

Press $\boxed{x/y}$ = and it should show 1.5

Brackets need to be interpreted carefully.

example

Calculate: **a** $\dfrac{240}{7 + 8}$ **b** $16 + 3(4 + 7)$ **c** $60 \div [30 - (13 + 5)]$

a $\dfrac{240}{(7 + 8)}$

$= \dfrac{240}{15} = 16$

b $16 + 3(4 + 7)$
$= 16 + 3 \times 11$
$= 49$

c $= 60 \div [30 - (13 + 5)]$
$= 60 \div [30 - 18]$
$= 60 \div 12$
$= 5$

Exercise N4.3

1 Calculate:
  a $12 + 6 \times 19$
  b $(1.1 + 3.9) \times (2.8 + 8.2)$
  c $8.4 - 0.4 \times 6$
  d $24 \div 4 \times 6$
  e $3.8 - 1.4 - 0.4$

2 Solve each of these calculations, using the most appropriate method.
  a $(18.7 - 1.3) \times (18.7 + 1.3)$
  b $6(7 + 5)$ **c** $\dfrac{7^2}{(3.5 \times 4)^2}$
  d $\dfrac{7 + 6}{3 \times 17}$ **e** $\dfrac{(4 + 5)^2}{\sqrt{(25 + 7 \times 8)}}$
  f $18 \div 4 \div 3$ **g** $4(5 + 13) + 6$
  h $120 \div [35 - (3 - 8)]$
  i $\sqrt{(21^2 - 14^2)}$

3 **Puzzle**
Kate and Simon are playing a game of Countdown. Simon explains how he got the target number of 114 from these numbers:

Target = 114

7	8	20
12	15	50

'First I multiplied 8 by 15, that makes 120.
Then I divided 120 by 20 to make 6.
Then I divided 12 by 6 to make 2.
50 add 7 is 57
... and $57 \times 2 = 114$.'
  a Write down Simon's calculation using the order of operations.
  b Use Simon's numbers to make a target number of 630. Write your answer as a calculation using the order of operations.

4 **Puzzle**
Insert the correct operations to make these calculations correct:
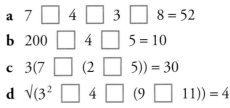
  a $7 \square 4 \square 3 \square 8 = 52$
  b $200 \square 4 \square 5 = 10$
  c $3(7 \square (2 \square 5)) = 30$
  d $\sqrt{(3^2 \square 4 \square (9 \square 11))} = 4$

5 Use your calculator to solve these problems:
  a From 8.15 am, Tom spends:
23 minutes travelling to school,
6 hours 37 minutes at school,
28 minutes travelling home and
2 hours 24 minutes watching TV.
He then has his evening meal. At what time does he have his evening meal?
  b Three consecutive whole numbers have a product of 50 616. What are the three numbers?
  c The area of a square pond is 14.2 m^2. What is the perimeter of the pond?

6 Solve each of these problems using the most appropriate method:
  a $2(3 + 6)^2 + 12$
  b $(^-18 + 2) \times (13 - {}^-17)$
  c $\dfrac{(1.2 \times {}^-5)^2 + 75}{\sqrt{(5 + 1.7 \times 3)}}$
  d $\dfrac{(6.8 - 2.2)^2}{\sqrt{(5.1 + 0.4 \times 8)}} + 4.7 \times 0.9$

7 **a** Copy and complete this table of values for the equation $y = 4x^2 + 12$.

x	$^-3$	$^-2$	$^-1$	0	1	2	3
$y = 4x^2 + 12$							

  b Does the point $(5, 115)$ lie on the line $y = 4x^2 + 12$? Explain your answer.

This spread will show you how to:

▶▶ Use standard column procedures for multiplication of decimals.

▶▶ Understand where to place the decimal point by considering equivalent calculations.

▶▶ Make and justify estimates and approximations of calculations.

KEYWORDS

Equivalent   Estimate

Multiplication

Decimal number

You can multiply decimal numbers using the standard method.
First you work out equivalent calculations.

example

Calculate 27.4×9.8

Estimate: $27.4 \times 9.8 \approx 27 \times 10 = 270$

Write an equivalent calculation: $27.4 = 27\frac{4}{10} = \dfrac{274}{10}$ $9.8 = 9\frac{8}{10} = \dfrac{98}{10}$

$$27.4 \times 9.8 = \frac{274}{10} \times \frac{98}{10} = \frac{274 \times 98}{100}$$

Use the standard method:

$$
\begin{array}{r}
274 \\
\times\ 98 \\
\hline
24660 \\
2192 \\
\hline
26852
\end{array}
$$

274×90

274×8

So $27.4 \times 9.8 = 274 \times 98 \div 100$

$= 26\ 852 \div 100$

$= 268.52$

You can estimate the size of large numbers by estimating a
small proportion and then multiplying by a scale factor.

example

In every 1 m² of his lawn, Bobby estimates there are 40 clover plants.
His rectangular shaped lawn measures 8 m by 5.4 m.
Estimate the number of clover plants on Bobby's lawn.

Area of lawn = $8 \times 5.4 = 43.2$ m²

Total number of clover plants = 40×43.2

$\approx 40 \times 40$

≈ 1600

8 m

5.4 m

Exercise N4.4

1 Calculate mentally:

  a 8×0.1 **b** 7×0.4
  c 11×8 **d** 12×15
  e 23×13 **f** 1.7×21
  g 0.8×12 **h** 29×18
  i 7.5×40 **j** 2.8×29
  k 6.5×140 **l** 42×30

2 **a** What is the cost of 34 loaves of bread at £0.79 each?

  b A plan of house has a scale of 1 : 16. The length of the main bedroom is 26 cm on the plan. What is the real length of the main bedroom?

  c What is the cost of 16 kg of cheese at £4.30 per kilogram?

3 Here are some facts:

World Records 2002 (men)		
100 metres	9.79 secs	Maurice Green
Long jump	8.95 m	Mike Powell
Javelin	98.48 m	Jan Zelezney

  a How long would it take Maurice Green to run the 1500 m if he could maintain his 100 m speed? In what time would he run the 10 000 m?

  b How far would Mike Powell jump if he could repeat his world record jump 23 times?

  c If Jan Zelezney could throw his world record throw every time, how far would he have thrown the javelin after 6 throws?

4 Calculate using the most appropriate method:

  a 0.4×0.5 **b** 3.74×8
  c 9.32×5 **d** 16.3×0.4
  e 31.4×4.5 **f** 17.1×3.6
  g 26.9×3.3 **h** 29.2×8.7
  i 12×4.31 m   **j** 23.8 m $\times 9.1$ m

5 **a** Coffee beans cost £12.30 per kg. How much do 7.4 kg of coffee beans cost?

  b Sebastian says that the height of a person is about 3.1 times the length of their arm. How tall is someone with an arm length of 58.7 cm?

  c Curtain fabric costs £3.70 per metre length. How much would a length of 2.8 m cost?

6 **Investigation**

Jonathon says 'When you multiply something by a number, you always make it bigger. When you divide something by a number, you always make it smaller.' Janet disagrees.

  a Copy and complete this multiplication table.

×	-2	-1.5	-1	-0.5	0	0.5	1	1.5	2
-2									
-1.5									
-1									
-0.5									
0									
0.5									
1									
1.5									
2									

  b Explain why Janet disagrees and show clearly for what numbers Jonathon's statement is correct.

7 **Puzzle**

Paula is running in a 10 km race.

  ▶ She runs the first lap in 72.1 seconds.
  ▶ For the next 12 laps she runs each 400 m in 71.3 seconds.
  ▶ For the next 7 laps she runs each 400 m in 73.4 seconds.

The world record for 10 km is 30 minutes.

  a How many laps of 400 m is 10 km?

  b In how many seconds must Paula run each of the remaining laps to equal the world record?

This spread will show you how to:
▶▶ Use standard column procedures for division of decimals.
▶▶ Understand where to place the decimal point by considering equivalent calculations.
▶▶ Make and justify estimates and approximations of calculations.
▶▶ Check a result by considering whether it is of the right order of magnitude and by working the problem backwards.

KEYWORDS

Division   Estimate
Equivalent   Calculate
Decimal number
Order of magnitude

You can use a written method based upon repeated subtraction to divide an integer by a decimal.

example

Calculate   **a** $494 \div 5.2$ **b** $183 \div 6.2$ to 1 decimal place

a   Estimate: $494 \div 5.2 \approx 500 \div 5 = 100$

Write an equivalent calculation:

$$494 \div 5.2 = \frac{494}{5.2} = \frac{4940}{52}$$

Now divide:

$$
\begin{array}{r}
52 \,) \, 4940 \\
- \, 4680 \quad 52 \times 90 = 4680 \\
\hline
260 \\
- \, 260 \quad 52 \times 5 = 260 \\
\hline
0
\end{array}
$$

So $494 \div 5.2 = 95$
Check by multiplying:

$$95 \times 5.2 = 494$$

b   Estimate: $183 \div 6.2 \approx 180 \div 6 = 30$

Write an equivalent calculation:

$$183 \div 6.2 = \frac{183}{6.2} = \frac{1830}{62}$$

Now divide:

$$
\begin{array}{r}
62 \,) \, 1830.00 \\
- \, 1240.00 \quad 62 \times 20 = 1240 \\
\hline
590.00 \\
- \, 558.00 \quad 62 \times 9 = 558 \\
\hline
32.00 \\
- \, 31.00 \quad 62 \times 0.5 = 31 \\
\hline
1.00 \\
- \, 0.62 \quad 62 \times 0.01 = 0.62 \\
\hline
0.38
\end{array}
$$

So $183 \div 6.2 = 29.5$ (1 dp)

Check by multiplying:

$$29.5 \times 0.62 = 182.9$$

The answer is not exactly the same, as 29.5 is rounded.

You should always estimate before calculating.
A good way to do this is to use approximate numbers.
The answer should be of the right order of magnitude.

A good approximation for $(18.7 - 4.6) \div (12.1 + 9.6)$
     is $(20 - 5) \div (10 + 10) = 15 \div 20 \approx 0.75$

A good approximation is easy to use and accurate.

Exercise N4.5

1 Calculate using an appropriate method:
   a 0.5 × 7   b 0.08 × 6
   c 0.04 × 10   d 1.2 × 9
   e 6.4 ÷ 0.01   f 644 ÷ 28
   g 770 ÷ 32   h 1070 ÷ 18

2 a 28 packets of sugar cost £22.40.
     What is the price of a packet of sugar?
   b The fastest talker in the world can
     speak at 627 words per minute. How
     many words does he speak in a second?
   c The maximum load for a lift is 960 kg.
     An average person weighs 75 kg.
     What is the maximum number of
     people that should travel in the lift?
   d Sharon has to cut 88.3 m of ribbon
     into 14 equally sized pieces. How long
     will each piece be (to 1 dp)?

3 a Copy and complete Betty's shopping
     bill:

Cost per item	Number of items	Total cost
	9	£21.06
£0.70		£9.10
	44	£5.28
	5	£18.45
£4.99	2	
£1.40		£21.00
£1.09		£3.27
	TOTAL	

   b How many items did Betty buy?

4 Calculate, giving your answer to 1 dp
   where appropriate.
   a 405 ÷ 1.5   b 615 ÷ 8.2
   c 81.4 ÷ 23   d 432 ÷ 6.4
   e 214 m ÷ 3.8   f 578 ÷ 5.8
   g 128 m ÷ 2.8 m h 189 kg ÷ 8.3

5 a In this 'productogon' the numbers in
     the squares are the products of the
     numbers on each side. Find the
     missing numbers.

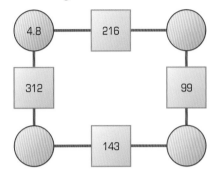

   b How would you solve this problem if
     you had not been given the first
     number of 4.8?

6 Here is some nutritional information
   from a 200 g packet of Mushroom Cous
   Cous:

Nutritional Information:

Composition (dry)	100 g provides
Energy	333 kcal
Protein	11.3 g
Carbohydrate	66.2 g
Fat	2.6 g
Fibre	4.9 g

A 50 g serving of dry Cous Cous weighs 75 g when cooked
Suitable for vegetarians.

ONLY £1.59!!

   a What is the cost of 12 packets of
     Mushroom Cous Cous?
   b How much dry Mushroom Cous Cous
     would you need to cook to provide
     you with 100 g of protein?
   c How many packets of Mushroom
     Cous Cous would you need to buy to
     provide you with 100 g of fibre?
   d How much energy is there in 120 g of
     cooked Mushroom Cous Cous?

This spread will show you how to:
- ▶▶ Use units of measurement to estimate, calculate and solve problems in everyday contexts.
- ▶▶ Enter numbers and interpret the display of a calculator in different contexts.

KEYWORDS
Convert   Hectare
Power
Tonne

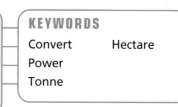

The metric system was designed to make it easy to convert between units by multiplying or dividing by powers of ten.

You should know the relationships between some common metric units.

Length	Weight	Capacity
1 km = 1000 m	1 tonne = 1000 kg	$1 m^3 = 1000$ litres
1 m = 100 cm	1 kg = 1000 g	1 litre = 1000 cm^3
1 cm = 10 mm	1 g = 1000 mg	1 ml = 1 cm^3

example

Convert 4.57 m into mm.

First convert to cm:

$$1 m = 100 cm$$

so   4.57 m = (4.57×100) cm   = 457 cm

× 100
Number of metres ⟶ Number of centimetres
÷ 100

Now convert to mm:

$$1 cm = 10 mm$$

so   457 cm = (457×10) mm   = 4570 mm

× 10
Number of centimetres ⟶ Number of millimetres
÷ 10

To convert between units of area, you need to know the relationships between the different units of area.

$$10\ 000\ m^2 = 1\ hectare$$

so $1200\ m^2 = (1200 \div 10\ 000)$ hectares = 0.12 hectares

1 square metre
$= 1\ m^2$
$= 100\ cm \times 100\ cm$
$= 10\ 000\ cm^2$

100 cm

100 cm

▶ Time does not use the decimal system.

To work out time on your calculator, you need to convert the decimal part.

60 seconds = 1 minute	60 minutes = 1 hour	24 hours = 1 day

453 seconds = $(453 \div 60)$ minutes = 7.55 minutes
= 7 minutes and 0.55×60 seconds
= 7 minutes and 33 seconds

÷ 60
seconds ⟶ minutes
× 60

Exercise N4.6

1 Convert these measurements into the units indicated:
  a 12 m = ____ cm
  b 8 kg = ____ g
  c 3 litres = ____ ml
  d 3.2 cm = ____ mm
  e 3 weeks = ____ days
  f 340 seconds = ____ minutes and seconds

2 In Crazyworld all road signs show measurements in millimetres.
  a Convert these distances into more sensible units:
    i 48 500 mm
    ii 370 000 mm
    iii 5 870 200 mm
  b Explain why measuring road distances in millimetres is not very sensible.

3 **a** Write down the mass of each of these objects:

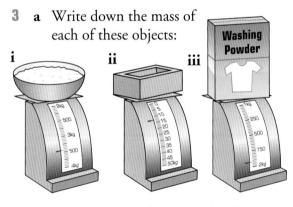

i   ii   iii

  b What is the total mass of the three objects?

4 **a** Jim arrived at 12.34 on the train from London. His journey had lasted 217 minutes. At what time did he start his journey?
  b Calculate the number of hours in 20 000 seconds.
  c Calculate the number of minutes you have been at secondary school since the beginning of Year 7.
  d Change 45 200 m^2 into hectares.

5 **Puzzle**
Johann is a football referee. Each year he has to complete the following tasks:
  ▸ Referee 48 football games (each lasting 125 minutes)
  ▸ Attend 7 training courses (each lasting 200 minutes)
  ▸ Attend 4 FA disciplinary hearings (each lasting 350 minutes)
  ▸ Attend 52 press conferences/interviews (each lasting 17 minutes)
  ▸ Inspect 12 football grounds (each lasting 215 minutes)
  a How many days does Johann work in a year as a referee?
  b If he gets paid £48 an hour, what is his annual salary?

6 Convert these measurements into the units indicated:
  a 17 mm into m
  b 1835 g into tonnes
  c 2.45 litres into cm^3
  d 24 000 mm into km
  e 3430 minutes into days, hours and minutes
  f 0.0048 hectares into mm^2

> **Hint:** 1 cm^2 = 100 mm^2

7 Calculate the perimeter (cm) and area (cm^2) of these shapes:

  a

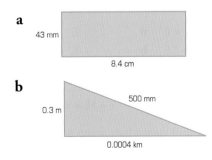

  43 mm

  8.4 cm

  b

  500 mm

  0.3 m

  0.0004 km

This spread will show you how to:
▶▶ Simplify or transform linear expressions by collecting like terms.
▶▶ Multiply a single term over a bracket.

KEYWORDS
Expression   Area
Factorise   Equivalent
Perimeter
Collect like terms

You can expand brackets using a multiplication grid.
This can be tricky when there are negative terms.

$6(7y - 3)$ ➡️

×	7y	⁻3
6	42y	⁻18

so $6(7y - 3) = 42y - 18$

$^-3(4p + 8)$ ➡️

×	4p	+8
⁻3	⁻12p	⁻24

so $^-3(4p + 8) = {}^-12p - 24$

$^-7(4 - 7r)$ ➡️

×	4	⁻7r
⁻7	⁻28	+49r

so $^-7(4 - 7r) = 49r - 28$

▶ To expand brackets you multiply each term inside the
  bracket by the term outside.

  $a(b + c) = ab + ac$

You may need to expand brackets to simplify expressions.

example

Simplify these expressions:

a $18 - 4(2x - 3)$

b $5(4 - 3x) - 3(5 - 2x)$

---

a   Expand the brackets:
       $18 - 8x + 12$ ⁻4 × ⁻3 = 12
       Collect like terms:
       $18 + 12 - 8x$
       Simplify:
       $30 - 8x$

b   Expand the brackets:
       $20 - 15x - 15 + 6x$ ⁻3 × ⁻2x = 6x
       Collect like terms:
       $20 - 15 - 15x + 6x$
       Simplify:
       $5 - 9x$

The expression $30 - 8x$ has two terms:
30 and ^-8x.
They have a common factor of 2.

You can write $30 - 8x = 2(15 - 4x)$.
This is called factorising the expression.

Exercise A5.1

1 Simplify these expressions:
- **a** $3x + 8 - 7x + 4$
- **b** $7x + 3 - 2x$
- **c** $7x - 5 + 13 - 2x$
- **d** $3y - 5 + y + 7$
- **e** $3x - 2 - 5x + 2$
- **f** $7x + 3y + 3x - 4y$

2 Copy and complete these addition grids:

a

+	2x − 3	5x + 4
6 − x		
2x − 7		

b

+	x + 7	
2x + 3		5x − 2
	3x − 2	

c

+		x − 3
	6x + 2	2x + 7
	3x − 5	

3 One of the missing expressions in each of these addition squares is $4 - 2x$. Find all of the expressions.

a

+	3x + 1	
2x − 5		3x + 2

b

+	3 + 2x	
		5 + 3x
	8 + 6x	

4 Simplify these expressions:
- **a** $3x + 2(x - 1)$
- **b** $2(3x - 1) + 5$
- **c** $2 - 3(x - 1)$
- **d** $2(3x - 2) + 5(x + 2)$
- **e** $3(2 - 3x) - 2(3 + 4x)$
- **f** $4(3 + 2x) - 2(5 - 3x)$

5 Factorise these expressions. The first two have been started for you.
- **a** $8x + 6y = 2(4x \quad)$
- **b** $15 - 10y = 5(\quad)$
- **c** $12y - 9x$
- **d** $18x - 12y$
- **e** $16 - 12x$
- **f** $x^2 - 2x$
- **g** $2x^2 + xy$
- **h** $5x^2 - 15x$

6 a Find the area shaded. Show all your working.

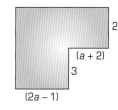

b Find the perimeter of the shape. Factorise your answer if possible.

7 Here are seven expression cards.

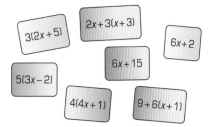

a Find three of the cards which are equivalent.
b Find a pair of cards where one is always thirteen larger than the other.
c For the bottom two cards ($4(4x + 1)$, $9 + 6(x + 1)$), what is their difference?
d Add all of the seven expressions together.
This total expression is equal to 20. What is the value of x?

This spread will show you how to:

▶▶ Construct and solve linear equations with integer coefficients with the unknown on both sides, using appropriate techniques.

KEYWORDS

Equation   Solve

Expression

Equivalent

In an equation the equals sign shows that the sides balance:

$$6p - 7 \quad 3p + 5$$

$$6p - 7 = 3p + 5$$

To solve an equation you must always keep the balance.

Take $3p$ from both sides:

$$6p - 3p - 7 \quad 3p - 3p + 5$$

$$6p - 3p - 7 = 3p - 3p + 5$$
$$3p - 7 = 5$$

You want all the p terms on one side of the equation and all the number terms on the other side.

The left-hand side has the most unknowns: $6p$ is more than $3p$, so you collect the p terms on the left.

Add 7 to both sides:

$$3p - 7 + 7 \quad 5 + 7$$

$$3p - 7 + 7 = 5 + 7$$
$$3p = 12$$

Divide both sides by 3:

$$3p \div 3 \quad 12 \div 3$$

$$3p \div 3 = 12 \div 3$$
$$p = 4$$

You may need to form equations to help solve problems.

Follow the instructions on each leg to complete this spider map.

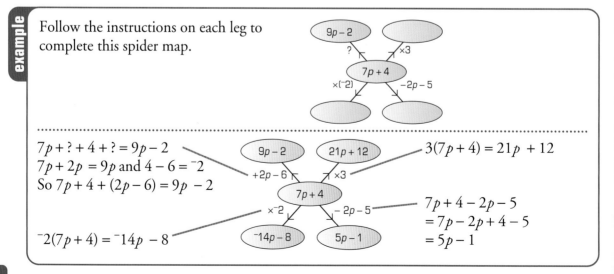

$7p + ? + 4 + ? = 9p - 2$
$7p + 2p = 9p$ and $4 - 6 = {}^-2$
So $7p + 4 + (2p - 6) = 9p - 2$

$3(7p + 4) = 21p + 12$

${}^-2(7p + 4) = {}^-14p - 8$

$7p + 4 - 2p - 5$
$= 7p - 2p + 4 - 5$
$= 5p - 1$

Exercise A5.2

1 Solve these equations:
  a $x + 2 = 17$
  b $y - 1 = 34$
  c $2x - 3 = 21$
  d $4x - 3 = 17$
  e $9 - 2x = 11$
  f $2x + 1 = 3x$

2 Copy and complete these spider maps.

a

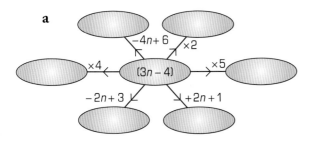

b

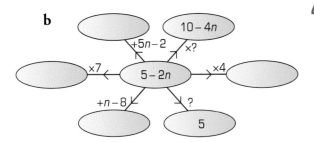

c

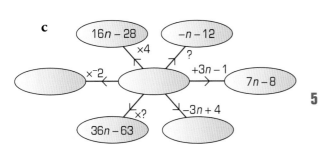

3 Each of these diagrams show two expressions that balance.
Write each balance as an equation and solve it.

a

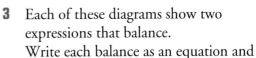

b

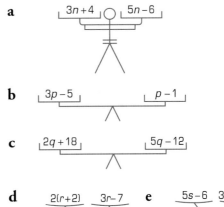

    $3p - 5$ $p - 1$

c $2q + 18$ $5q - 12$

d $2(r+2)$ $3r - 7$ **e** $5s - 6$ $3(9 - 2s)$

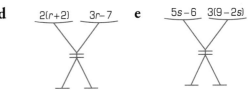

4 a Solve each of these equations:
    i $6a + 1 = 2a + 25$
    ii $2e + 8 = 4e - 12$
    iii $4e + 2 = 5e + 4$
    iv $2i + 7 = 5i - 2$
    v $6 + 3l = l + 22$
    vi $20 - n = n - 10$
    vii $6q + 7 = 3q + 7$
    viii $4t - 40 = 2t + 10$
    ix $5u + 7 = 20 - 8u$
    x $18 - 2v = v + 3$

  b Order your answers from smallest to largest. The order of the letters will spell a word used in algebra. What is it?

5 Here are three expressions:
  A $2x + 25$ B $2(4x - 7)$ C $3(10 - x)$
  a Find x if A is equivalent to B.
  b For what values of x is A = C **and** B = C?
  c Find the value of x when 2A = B.

A5.3 Solving equations with brackets

This spread will show you how to:

▶▶ Construct and solve linear equations with integer coefficients with brackets, using appropriate techniques.

In an arithmagon, each number in a square is the sum of the numbers on either side of it.

You can use algebraic expressions in arithmagons. The rules are the same.

9

16   24

7 — 22 — 15

$9 + 15 = 24$

example

Find expressions in x for each of the missing values in this arithmagon.

3x – 4

A   B

7 – 2x — 10 + 3x — C

Square A $= (3x - 4) + (7 - 2x)$
$\quad = 3x - 2x - 4 + 7$
$\quad = x + 3$

Circle C $= (10 + 3x) - (7 - 2x)$
$\quad = 10 + 3x - 7 + 2x$
$\quad = 10 - 7 + 3x + 2x$
$\quad = 3 + 5x = 5x + 3$

Square B $= (3x - 4) + (5x + 3)$
$\quad = 3x + 5x - 4 + 3$
$\quad = 8x - 1$

You can solve equations with brackets using the balance method.

example

Solve $3(2d - 4) = 5(2 - d)$

	$3(2d - 4)$	$= 5(2 - d)$
Expand the brackets:	$6d - 12$	$= 10 - 5d$
Add $5d$ to both sides:	$6d + 5d - 12$	$= 10 - 5d + 5d$
	$11d - 12$	$= 10$
Add 12 to both sides:	$11d - 12 + 12$	$= 10 + 12$
	$11d$	$= 22$
Divide both sides by 11:	$11d \div 11$	$= 22 \div 11$
	d	$= 2$

Check your answer: $3(2d - 4) = 3(2 \times 2 - 4) = 3 \times 0 = 0$
$5(2 - d) = 5(2 - 2) = 5 \times 0 = 0$

They are the same so $d = 2$ is correct.

Exercise A5.3

1 a Copy and complete this arithmagon:

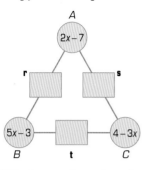

b Work out the values for 2A, 2B and 2C.

c Find 2A + 2B + 2C and **r** + **s** + **t**.

d Compare the two answers. Explain what you find.

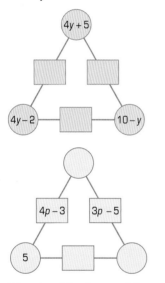

e Check with these arithmagons that your rule works.

2 Find the value for each unknown in these equations:

a $2(n + 5) = 3(n - 2)$

b $4(2m - 3) = 5(m + 3)$

c $3(4 - 2p) = 2(p - 6)$

d $5(3q + 2) = 4(4q + 1)$

e $4(7 - 3x) = 4$

f $5(3r + 8) - 7 = 3$

g $6(5t - 8) = 4(3t + 7) - 6$

h $3(8 - 3u) = 4u + 2(5 - 3u)$

3 Solve these equations using two different methods.
For example:

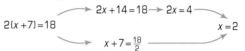

Show **two** methods to solve these equations:

a $5(x - 3) = 35$

b $4(2x + 7) = 100$

c $3(14 - 3x) = 15$

Choose the most appropriate method to solve these equations:

d $5(7x - 4) = 85$

e $4(5x + 2) = 68$

f $3(9 - x) = 18$

4 Solve these equations using an appropriate method.

a $5x = 13$ **b** $8 - 3x = 17$

c $\dfrac{x}{5} + 5 = 11$ **d** $4x - 5 = 24$

e $3x + 7 = 6x - 8$ **f** $7x + 9 = 3x + 3$

g $2(4x - 3) = 14$ **h** $\dfrac{5x}{3} = 7$

i $16 = \dfrac{3x}{4}$ **j** $\dfrac{(x + 14)}{3} = 3$

5 Challenge
The expression in a brick is the sum of the two expressions above it.
Fit in the expressions to make these two towers correct.

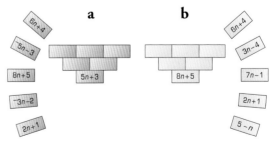

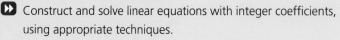

A5.4 Equations involving fractions

This spread will show you how to:

▶▶ Construct and solve linear equations with integer coefficients, using appropriate techniques.

KEYWORDS

Equation   Solution
Denominator Ascending

A fraction has a denominator which means you divide.

$\frac{3}{4}$ means $3 \div 4$.

Algebraic expressions may also include a fractional part.
You can undo the division by multiplying.

example

Solve $\dfrac{3t-4}{5} = 4$ to find the value of t.

You can think of the left-hand side as having a bracket: $\dfrac{(3t-4)}{5}$

> You can write the equation as $(3t-4) \div 5 = 4$

Multiply both sides by 5:

$$\frac{(3t-4)}{5} \times 5 = 4 \times 5$$
$$(3t-4) = 20$$

Add 4 to both sides:

$$3t - 4 + 4 = 20 + 4$$
$$3t = 24$$

You can also solve equations with algebraic denominators.

example

Solve $\dfrac{6}{(x-2)} = \dfrac{5}{(x-4)}$

You can cross multiply to get rid of the denominators:

$$\frac{6}{(x-2)} = \frac{5}{(x-4)}$$
$$6(x-4) = 5(x-2)$$

Expand the brackets:

$$6x - 24 = 5x - 10$$

Subtract $5x$ from both sides:

$$6x - 5x - 24 = 5x - 5x - 10$$
$$x - 24 = {}^-10$$

Add 24 to both sides:

$$x - 24 + 24 = {}^-10 + 24$$
$$x = 14$$

Exercise A5.4

1 Solve these questions by first removing the fractions:

a $\dfrac{x+5}{3} = 2$

b $\dfrac{x-7}{2} = 4$

c $\dfrac{2x+1}{5} = 3$

d $\dfrac{3x-5}{2} = 2$

e $\dfrac{5x+4}{3} = 8$

f $\dfrac{2x+1}{3} = 7$

2 Solve these further equations by first removing the fraction part.

a $\dfrac{2x+5}{3} = 7$

b $\dfrac{4x-3}{2} = 4\dfrac{1}{2}$

c $\dfrac{2x+7}{5} = 3$

d $\dfrac{3x-5}{2} = x+2$

e $\dfrac{5x+4}{3} = 3x-8$

f $\dfrac{2x+1}{3} = \dfrac{3x-2}{4}$

3 **Odd one out**
Which equation has a different solution to the other two?

$3x+5 = x+19$ $5(x-2) = 3x+4$ $\dfrac{7x+2}{3} = \dfrac{2x+13}{5}$

4 **Mixed bag**
   a Solve these equations:
   i $2 - 3a = 11$
   ii $2e + 10 = 0$
   iii $2(3k+4) = 80$
   iv $\dfrac{8}{o} = 2$
   v $2(5r+5) = 4(3r-1)$
   vi $\dfrac{5+3s}{2} = 1$
   vii $\dfrac{(2w+8)}{7} = w-1$
   viii $\dfrac{4y-3}{2} = \dfrac{y-6}{4}$

   b Put your solutions in ascending order. The letters used as the variable now spell two words. Are they true?

5 Solve these equations by first removing the fractions.

   a $\dfrac{1}{x} = 2$

   b $\dfrac{9}{2x} = 3$

   c $2 = \dfrac{6}{x+2}$

   d $\dfrac{x+4}{x} = 3$

This spread will show you how to:

▶▶ Begin to use graphs and set up equations to solve simple problems involving direct proportion.

KEYWORDS

Proportion   Graph
Steepness   Ratio
Equation

Jules and Saeed are mixing paint to try to find the perfect orange.
They mix yellow and red.

Equal quantities of yellow and red make this colour orange:

The graph shows the different amounts of each colour they can use to make the same colour.

If they use 5 litres of red, they must use 5 litres of yellow.

Proportion of yellow = Proportion of red.

You can write this as a ratio:
yellow : red
   = 1 : 1

The equation of the graph is $y = x$.

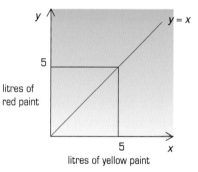

example

These graphs show the proportions of red and yellow paint used in three different mixtures.

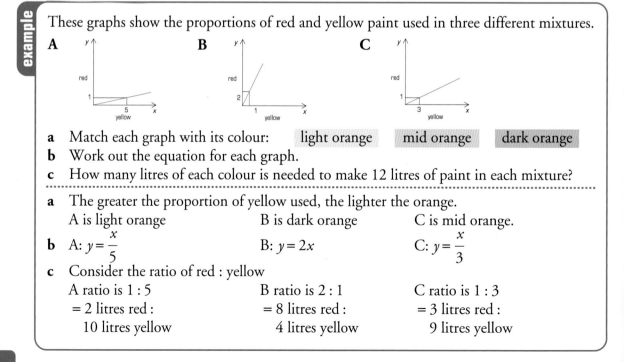

a   Match each graph with its colour: light orange   mid orange   dark orange
b   Work out the equation for each graph.
c   How many litres of each colour is needed to make 12 litres of paint in each mixture?

a   The greater the proportion of yellow used, the lighter the orange.
     A is light orange   B is dark orange   C is mid orange.

b   A: $y = \dfrac{x}{5}$ B: $y = 2x$ C: $y = \dfrac{x}{3}$

c   Consider the ratio of red : yellow
     A ratio is 1 : 5   B ratio is 2 : 1   C ratio is 1 : 3
     = 2 litres red :   = 8 litres red :   = 3 litres red :
       10 litres yellow   4 litres yellow   9 litres yellow

Exercise A5.5

1 Each of these graphs show how white and red paint are mixed in proportion to make shades of pink paint.

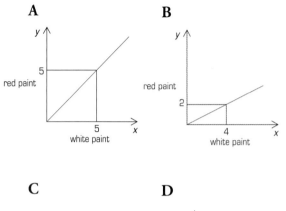

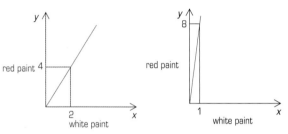

a In graph B how much red paint would you mix with 4 litres of white paint?

b In graph C how much white paint would you mix with 6 litres of red paint?

c Which graph shows the mixture that gives darkest pink paint?

d Which graph shows the mixture that gives the lightest pink paint?

e Match these equations with three of the graphs (y represents the amount of red paint, x the amount of white paint):
$$y = \tfrac{1}{2}x \qquad y = 8x \qquad y = x$$

f Write the equation for the fourth graph. (Check with numbers that it is correct.)

g Which graph is the steepest? How is this shown in the equations?

2 Purple paint is made by mixing red and blue paint.

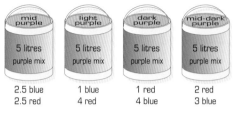

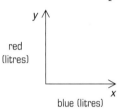

a On a grid like this sketch a line to show each mixture of paint.

b One graph has equation $3y = 2x$ or $y = \tfrac{2}{3}x$
What are the equations for the other three graphs?

c Do steep graphs mean a dark purple or a light purple?

3 Four shades of blue were made by mixing blue paint with different amounts of white paint.

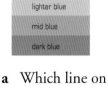

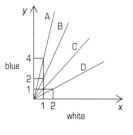

a Which line on the graph represents each blue?

b Work out the equation for each line.

c There were 15 litres of paint mixed in each colour.
Work out how many litres of blue and white were used in each case.

This spread will show you how to:

▶▶ Plot the graphs of linear functions, where x is given in terms of y.

▶▶ Recognise that graphs of the form $y = mx + c$ are straight lines.

KEYWORDS

Linear function

Graph   Intercept

Gradient

A linear function makes a straight-line graph.

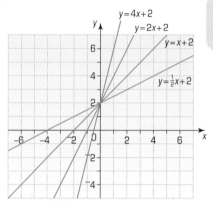

The equations are all of the form:
$y = mx + 2$
2 is the y-intercept.

These graphs all pass through the point $(0, 2)$.

The bigger the multiplier, the steeper the graph.

▶ The equation of a linear graph is of the form:

$$y = mx + c$$

y-coordinate   gradient x-coordinate y-intercept

You measure the gradient by choosing two points on the line:

▶ Gradient $= \dfrac{\text{vertical distance between the points}}{\text{horizontal distance between the points}}$

Find the gradient, y-intercept and equation for each of these lines:

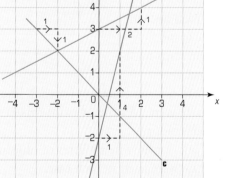

a $m = \frac{4}{1} = 4$ and $c = {}^-2$.
Equation is $y = 4x - 2$

b $m = \frac{1}{2}$ and $c = 3$.
Equation is $y = \frac{1}{2}x + 3$

c $m = \frac{-1}{1} = 1$ and $c = 0$.
Equation is $y = {}^- x$

Exercise A5.6

1 Match each statement in **a** to an equation in **b** and its graph in **c**.

a

A gradient = 2	B gradient = 1
C gradient = ⁻2	D gradient = $\frac{1}{2}$

b

E $y = 3 - 2x$	F $y = \frac{1}{2}x + 2$
G $y = x + 2$	H $y = 2x - 2$

c

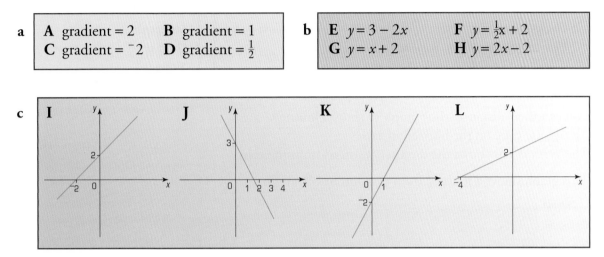

2 For each of these graphs work out:
  a The gradient   **b** The intercept   **c** The equation of the graph.

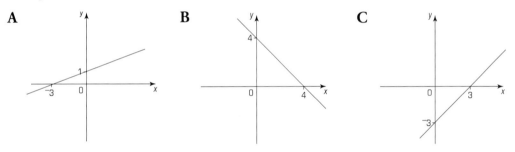

3 On the same axes draw the graph of $y = 2x + 1$ and $y = 7 - x$.
Write down the coordinate pair where the graphs intersect.

4 Kabir's homework was to draw out a table of coordinate pairs and
draw the graph of $y = 2x - 3$.
He has made a number of
mistakes.
Find the mistakes and draw the graph correctly.
Explain where you think Kabir went wrong.

x	⁻3	⁻2	⁻1	0	1	2	3
y	⁻3	⁻1	⁻1	⁻3	⁻1	1	3

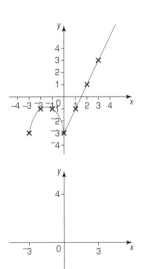

5 Quadratic challenge
The graph of $y = x^2$ is not a
straight line graph.

x	⁻3	⁻2	⁻1	0	1	2	3
$y = x^2$		4					

  a Copy and complete this table of values.
  b Plot the coordinate pairs on a grid like the one shown. Join these
     up to make a smooth curve.
  c Use your graph to find the value of y when $x = 2.5$

This spread will show you how to:
▶▶ Construct linear functions arising from real-life problems and plot their corresponding graphs.
▶▶ Discuss and interpret graphs arising from real situations.

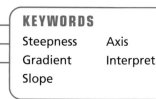

You can use a distance–time graph to show a journey.
Always use time on the horizontal axis.
The vertical axis shows the distance travelled from the start.

This graph shows a journey from London to Birmingham.

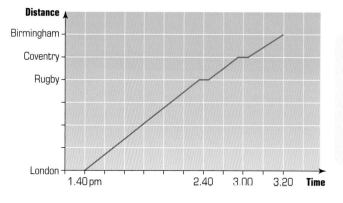

▶ A horizontal line on the graph shows no movement. You stay in the same place for this period of time.

▶ The sloped parts show movement. The steeper the slope, the faster the movement.

example

The distance–time graph shows a train journey from Cardiff to London.

a At what time did the train leave Cardiff?
b How many stations did the train stop at on the way?
c What was the total time taken for the journey?
d What was the fastest part of the journey?

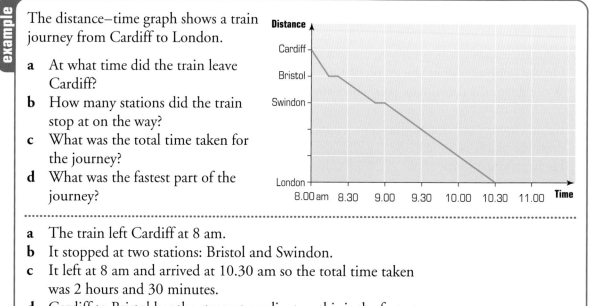

a The train left Cardiff at 8 am.
b It stopped at two stations: Bristol and Swindon.
c It left at 8 am and arrived at 10.30 am so the total time taken was 2 hours and 30 minutes.
d Cardiff to Bristol has the steepest gradient so this is the fastest part.

Exercise A5.7

1 This graph shows a journey in a lift.

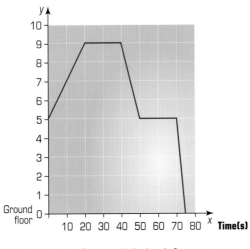

  a At which floors did the lift stop?
  b How long was the total journey?
  c Describe the journey in words.

2 This graph shows Kim's journey to school, after getting up in the morning.

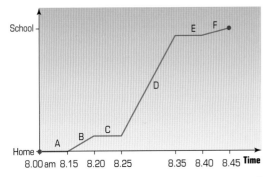

These six statements explain each stage of Kim's journey to school. Match the statements with the stages on the graph.
  a Travel by bus
  b Walk to the bus stop from home
  c Eat quick breakfast
  d Walk to school from the bus
  e Wait for the bus to arrive
  f Stop at shop to buy a drink.

3 This graph shows the height of water in the bowl while Jeff did the washing up.

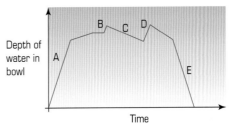

Interpret what may be happening at A, B, C, D and E.

4 Class 8T go swimming in town.
This is the timetable of events:
2.00 Bus leaves school for swimming pool
2.15 Arrive at swimming pool
3.15 Bus leaves swimming pool for school
3.25 Arrive at school

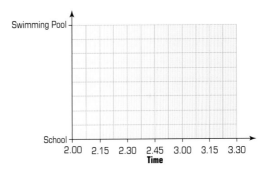

  a Copy and complete the distance–time graph to show this information.
  b How did you show the time at the pool?

5 Boiling water is poured into three metal cylinders.
  ▶ Cylinder A is left on the bench top.
  ▶ Cylinder B is placed in a bucket of ice.
  ▶ Cylinder C is wrapped in insulation.
The temperature is recorded over an hour.
Sketch a temperature–time graph for each cylinder.

This spread will show you how to:

▶▶ Simplify or transform linear expressions by collecting like terms.

▶▶ Construct and solve linear equations.

KEYWORDS

Equation   Isosceles

Equilateral   Perimeter

Solve

This diagram shows a lot of information.

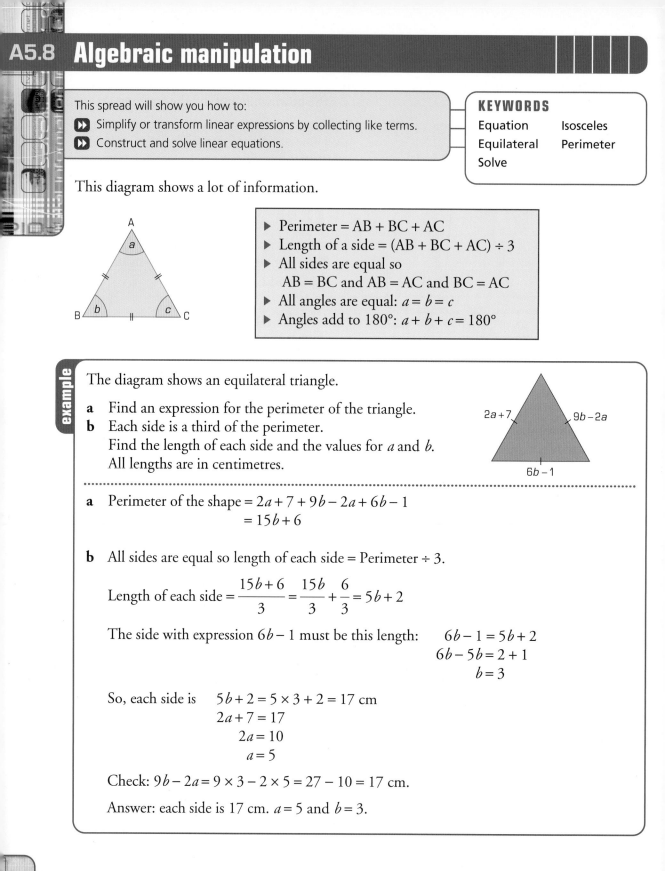

▶ Perimeter = AB + BC + AC

▶ Length of a side = (AB + BC + AC) ÷ 3

▶ All sides are equal so
  AB = BC and AB = AC and BC = AC

▶ All angles are equal: $a = b = c$

▶ Angles add to 180°: $a + b + c = 180°$

example

The diagram shows an equilateral triangle.

a Find an expression for the perimeter of the triangle.

b Each side is a third of the perimeter.
Find the length of each side and the values for a and b.
All lengths are in centimetres.

$2a + 7$ $9b - 2a$

$6b - 1$

a Perimeter of the shape = $2a + 7 + 9b - 2a + 6b - 1$
$$= 15b + 6$$

b All sides are equal so length of each side = Perimeter ÷ 3.

Length of each side $= \dfrac{15b + 6}{3} = \dfrac{15b}{3} + \dfrac{6}{3} = 5b + 2$

The side with expression $6b - 1$ must be this length:
$$6b - 1 = 5b + 2$$
$$6b - 5b = 2 + 1$$
$$b = 3$$

So, each side is $5b + 2 = 5 \times 3 + 2 = 17$ cm
$$2a + 7 = 17$$
$$2a = 10$$
$$a = 5$$

Check: $9b - 2a = 9 \times 3 - 2 \times 5 = 27 - 10 = 17$ cm.

Answer: each side is 17 cm. $a = 5$ and $b = 3$.

Exercise A5.8

1 Each of these are equilateral triangles:
   i Find the perimeter for each in terms of a and b.
   ii Find the actual length of each side and the values for a and b.

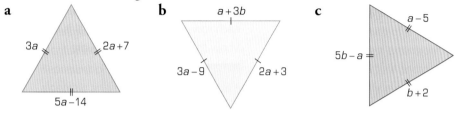

a $3a$, $2a+7$, $5a-14$

b $a+3b$, $3a-9$, $2a+3$

c $a-5$, $5b-a$, $b+2$

2 Each of these triangles is isosceles. Use this fact to make an equation to solve the problems.

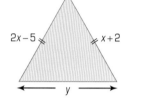

a $2x-5$, $x+2$, y

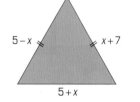

b $5-x$, $x+7$, $5+x$

c $2x+3$, $5x-6$, $10+x$

If the perimeter is 26 cm, what is the value of y?

By finding the value for x, find the perimeter of this triangle.

In this isosceles triangle you do not know the equal sides. Make three equations to find the possible values of x.

3 a Draw a square with the same perimeter as this rectangle. What is the area of the square?
   b i Find the perimeter of the square with the same area as this rectangle.
   ii Another square has the same perimeter as this rectangle. Sketch this square and find its area.

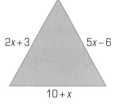

14 cm, 2 cm

9 cm, 4 cm

4 In this right-angled triangle the height is twice its base. What is the value of x?

5 Challenge
   In this right-angled triangle, what are the lengths of the sides:
   a If it is isosceles
   b If the height is twice the length of the base
   c If the height is three times the length of the base?
   d Show why the height cannot be four times the length of the base.

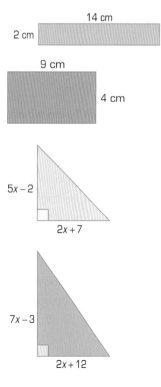

$5x-2$, $2x+7$

$7x-3$, $2x+12$

Organising your results

This spread will show you how to:
- ▶▶ Represent problems and interpret solutions in graphical form.
- ▶▶ Solve more complex problems by breaking them into smaller steps or tasks.

KEYWORDS
Interpret
Independent variable
Dependent variable

Using tables and graphs is a good way of organising your results to help you spot trends or patterns.

The table and graph show the connection between the number of red and yellow slabs in a series of hexagon paths.

The table makes it easy to read off particular values.

The graph emphasises the fact that the number of yellow slabs goes up steadily as the number of red slabs increases.

Red slabs	Yellow slabs
1	6
2	10
3	14
4	18
5	22
6	26

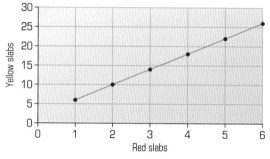

The line between the points does not represent any actual arrangement but is useful for showing the trend.

▶ Use tables to organise data, identify patterns and spot errors.

▶ Graphs show you a picture of your data.
You can view a lot of information at the same time, see trends and spot errors without having to read the figures.

The 'number of red slabs' and 'number of yellow slabs' can take different values – they are variables.

You are free to choose the number of red slabs: independent variable.
The 'number of yellow slabs' depends on the number of red slabs: dependent variable.

▶ The **independent** variable is plotted on the **horizontal** (x) axis.
▶ The **dependent** variable is plotted on the **vertical** (y) axis.

Exercise P1.2

1 **a** Draw a graph to illustrate the information from the table you made in question **2** on page 195. Your graph should show how many yellow slabs are needed for paths with up to 10 red slabs.

    b Make a new graph to show the length of edging strip needed for paths with up to 10 red slabs.

2 This table shows the number of slabs needed for the path design in question **6** on page 195.

    a Make a graph to show the number of yellow slabs needed for various paths of this design.

    b Make a graph to show the length of edging strip required.

Red slabs	Yellow slabs	Edging strip (cm)
1	4	160
2	6	220
3	8	280
4	10	340
5	12	400

3 This 1-bay bookshelf uses:
- ▶ 5 shelves
- ▶ 2 uprights
- ▶ 2 cross-pieces.

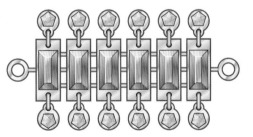

This 2-bay bookshelf uses:
- ▶ 10 shelves
- ▶ 3 uprights
- ▶ 4 cross-pieces.

    a Make a table to show how many shelves are needed for up to 10 bays.

    b Use the information from your table to produce a graph. Use the number of bays as the independent variable.

    c Repeat parts **a** and **b** for the number of uprights.

    d Make a table and graph to show the connection between the number of cross-pieces and the number of bays.

4 This bracelet is made of a number of sections. Each section contains one ruby and two emeralds. There is one silver ring at each end of the bracelet.

The graph shows:
- ▶ the number of silver rings
- ▶ the number of rubies
- ▶ the number of emeralds.

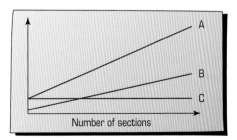

The independent variable is the number of sections in the bracelet.

Write a paragraph explaining what each line represents. Explain your reasoning fully.

Number of sections

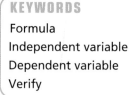

This spread will show you how to:

▶▶ Represent problems and interpret solutions in algebraic form, using correct notation.

▶▶ Use logical argument to establish the truth of a statement.

Formula
Independent variable
Dependent variable
Verify

Once you have an algebraic rule or formula you can predict any results.

This pattern is made of yellow rectangles surrounded by green strips.

The number of yellow rectangles (y) is the independent variable, and the number of green strips (g) is the dependent variable.

Your rule will start: $g=$

For every new yellow rectangle you need 3 new green strips.

The rule will be something like: $g = 3y$

You need one extra strip for the first pattern, so the rule will be:

$$g = 3y + 1$$

dependent   independent

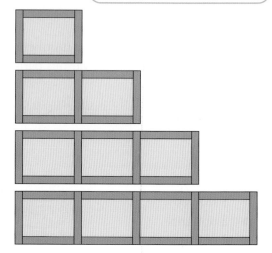

It can be easier to put the results in a table and then look for the rule:

y	1	2	3	4
g	4	7	10	13

+3   +3   +3

The value of g goes up by 3 every time the value of y goes up by 1.

So the rule will be similar to $g = 3y$.

y	1	2	3	4
$3y$	3	6	9	12

The values of g are 1 bigger than $3y$, so the formula is $g = 3y + 1$

example

Show that for 100 yellow squares you will need 301 green strips.

. . . .

Use the formula $g = 3y + 1$.
$y = 100$, so $g = 3 \times 100 + 1 = 301$

This is much easier than 'adding on 3' 100 times!

Exercise P1.3

1 Find a rule for each table. The first one is done for you.

a
x	y
1	5
2	8
3	11
4	14
5	17

$y = 3x + 2$

b
x	w
1	11
2	12
3	13
4	14
5	15

$w =$

c
x	z
1	7
2	12
3	17
4	22
5	27

$z =$

2 The first three patterns in a sequence are:

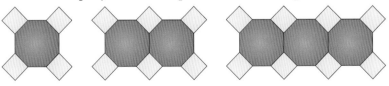

Pattern 1
2 green strips
2 black strips

Pattern 2
3 green strips
4 black strips

Pattern 3
4 green strips
6 black strips

a Find a rule for the number of green strips in any pattern.
Use the pattern number as the independent variable.
b Find a rule for the number of black strips in any pattern.

3 Using the table you made in question **2** on page 195:
a Find a rule for the number of yellow slabs in any path.
Use the number of red slabs as the independent variable.
b Find a rule for the length of edging strip required for any number of red slabs.

4 The Primrose Path Company also makes paths with this design.

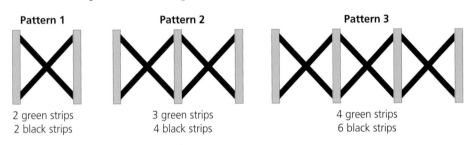

The length of each side of both shapes is 10 cm.
a Find a rule for the number of yellow slabs in any path.
b Find a rule for the length of edging strip required for paths with any number of red slabs.
c Show that for 100 red slabs you will need 202 yellow slabs and 6100 cm edging strip.

5 Question **3** on page 197 shows the design for a shelving system.
Find an algebraic rule for the number of:
a shelves   b uprights   c cross-pieces
needed for designs of any size. Use the number of bays as the independent variable.

P1.4   Investigating ratios

This spread will show you how to:

▶▶ Represent problems and interpret solutions in graphical form.

▶▶ Consolidate understanding of ratio and proportion.

▶▶ Use the unitary method to solve simple word problems.

KEYWORDS

Ratio   Proportion

Interpret

The Primrose Path Company want to know how many of each type of slab to make. They need to think about ratios.

The first three patterns in the 'Squares and Octagons' design are:

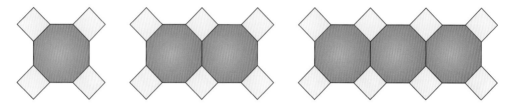

The table shows the ratio of yellow slabs to red slabs.

Pattern number	1	2	3
Red slabs	1	2	3
Yellow slabs	4	6	8
Ratio of yellow : red	**4 : 1**	6 : 2 = **3 : 1**	8 : 3 = **2.67 : 1**

The ratios are written as unitary ratios – just divide the first number by the second.

The graph illustrates the connection between the pattern number and the ratio of yellow slabs to red.

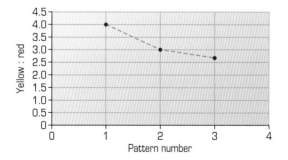

It is much easier to compare ratios when they are written as unitary ratios!

The line connecting the points shows the trend.

The company clearly need to make more yellow slabs than red ones. The graph shows that the longer the path, the smaller the proportion of yellow slabs needed in comparison to red slabs.

Exercise P1.4

1 The table shows the values of *a* and *b* and the unitary ratio of *b* to *a*.

a	1	2	3	4	5	6	7
b	9	14	19	24	29	34	39
Ratio b : a = b ÷ a	9						

Copy and complete the table.

2 Copy and complete the graph for the ratios in question **1**.

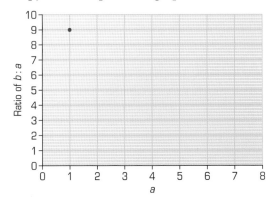

3 The Primrose Path Company want to know the ratio of yellow to red slabs for their Hexagons path design. Draw a table to show the ratio for Hexagon paths of various lengths.

4 Show the information from your table in question **3** on a graph. Use the number of red slabs as the independent variable, and the unitary ratio as the dependent variable.

5 Marcia's brainwave is to sell the 'Squares and Octagons' path in the form of a starter pack and some extension packs. The ratio yellow : red is shown below the packs.
  a Explain how the 'Hexagons' design could be sold in the same way.
  b Use part **a** to explain the shape of the graph in question **4**.

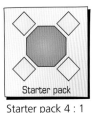

Starter pack
Starter pack 4 : 1

Extension pack
Extension pack 2 : 1

The longer the path, the closer the ratio will be to 2 : 1

6 a Repeat the steps in questions **3–5** using the information from question **3** on page 197.
    Use the number of bays as the independent variable and your own choice as the dependent.
  b Choose one of the other variables (e.g. the number of shelves), and work out the ratio of that variable to the number of bays.

This spread will show you how to:
- ▶▶ Break a complex problem into smaller steps or tasks, choosing and using efficient techniques for calculation.
- ▶▶ Suggest extensions to problems, conjecture and generalise.
- ▶▶ Design and use two-way tables.

KEYWORDS
Two-way table
Graph
Deduce

In any investigation, start with the simplest arrangement and move on to the more complicated ones.

Here is another extract from the Primrose Path Company's catalogue:

WIDER PATHS FOR THE BIGGER GARDEN

Treat yourself to the luxury of a Primrose Path in a W I D E R arrangement. Use our standard parts to build any of our designs in a double or treble width. Here are a few of the limitless possibilities.

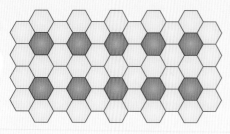

Part of a double-width path using our popular Hexagon range.

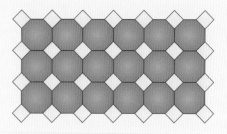

This is our stunning 'Squares and Octagons' design in a spectacular treble width.

Now the length and the width can change.
You have two independent variables: the number of red slabs across, and down.

You can organise the results into a two-way entry table:

Number of yellow slabs		Red slabs across				
		1	2	3	4	5
Red slabs down	1					
	2					
	3					
	4					
	5					

You can show the information on a graph using a separate line for each row.

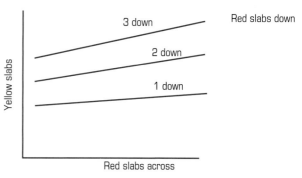

Exercise P1.5

1 **a** Copy and complete the two-way entry table on page 202,
to show the number of yellow slabs for paths of different
lengths and widths of the hexagons design.

   b Use the information in your table to make an accurate
version of the graph on page 202.

2 Design and complete a two-way entry table for the
'Squares and Octagons' design.
Your table should include designs with up to 5 red slabs across,
and up to 5 red slabs down.

3 Use the table you completed in question **2** to
produce a chart to show the same information.
Deduce what your graph shows.

4 A table mat is made of metal squares held together with wire rings.
This design is 4 squares long and 3 squares wide.
It uses 22 wire rings.

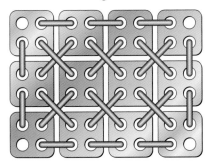

   a Design and complete a two-way entry table to show
how many wire rings are needed for mats up to
5 squares long and 5 squares wide.

   b Show the information from your table in a graph.

5 Start with a cuboid, one cube high.
Make a pyramid on this base by adding cuboids that are one cube high,
one cube shorter in width and length.
Find the volume of your pyramid.
For example:

Start with a 3 × 4 cuboid   Add a 2 × 3 cuboid   then a 1 × 2 cuboid

Investigate for other sizes of base, keeping your results in a two-way table.

This spread will show you how to:
- ▶▶ Use logical argument to establish the truth of a statement.
- ▶▶ Suggest extensions to problems, conjecture and generalise.
- ▶▶ Identify exceptional cases and counter-examples.

KEYWORDS

Two-way table   Rule
Exceptional case
Counter-example

When you look at an extension to a problem, you often need
to start the problem solving cycle again.

Here is the back page of the Primrose Path Company's catalogue.

Our kits provide the perfect solution for building an attractive
border around flowerbeds, ponds and pools.
Here are just a couple of examples; the possibilities are endless!

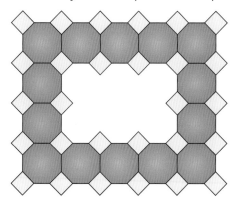
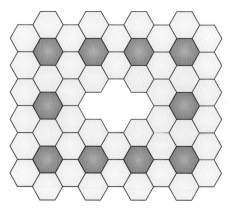

These are single-width borders, but double-width borders are
just as possible – or even wider!

Another great feature of our kits is that you don't have to stick
to straight lines. See below for a couple of examples of how you
can use curves to add interest or avoid obstacles.

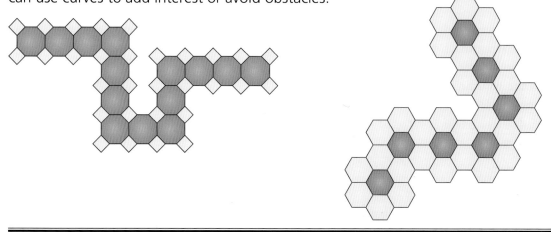

Exercise P1.6

1 Say whether these statements are true.
Explain your reasoning carefully.
  a All squares are rectangles.
  b All rectangles are squares.
  c Andy can eat one apple pie in 5 minutes.
     This means he can eat 24 apple pies in 2 hours.
  d If a, b, c, and d are any four numbers, then
     $(a + b) \times (c + d) = (a \times c) + (b \times d)$
  e For any set of two numbers, the mean and the median must be equal.

> An exceptional case is an example which breaks a general rule or hypothesis proving that the rule is not true.

2 Investigate different arrangements of paths and borders.
Here are a few suggestions ...

  a The first design shown on page 204 is '5 octagons across and 4 octagons down'.
     It needs 14 red slabs and 28 yellow ones.
     You could show these results in two-way entry tables.

	Octagons across			
Yellow slabs	3	4	5	6
Octagons down 3				
Octagons down 4			28	
Octagons down 5				
Octagons down 6				

	Octagons across			
Red slabs	3	4	5	6
Octagons down 3				
Octagons down 4			14	
Octagons down 5				
Octagons down 6				

  ▶ Copy and complete the tables to show the results for paths of different sizes.
  ▶ Describe any patterns you can see in the tables.
  ▶ Find algebraic rules to describe the patterns you find.

  b How do the rules change when the width of the path changes?
  c Investigate the ratio of yellow slabs to red ones for various arrangements.
  d Investigate paths that are not straight. Some arrangements follow the same rules as 'straight' paths – some don't. Can you find out which ones do, and which ones don't?

> A counter-example shows that a statement is not always true.

You should know how to ...

1 Identify the necessary information to solve a problem.

2 Represent problems and interpret solutions in algebraic and graphical form, using correct notation.

3 Use logical argument to establish the truth of a statement.

4 Use the unitary method to solve simple word problems.

Check out

1 This is the third pattern in a sequence made from red squares and yellow hexagons.
Explain how you would work out the relationship between the number of squares and the number of hexagons in a pattern.

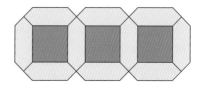

2 For the pattern shown in question **1**, use s to stand for the number of squares, and h for the number of hexagons. The formula connecting h and s is:

$$h = 3s + 1.$$

a Explain why this formula works.

b Draw a graph to show the relationship between h and s. Use s as the independent variable.

3 Sam looks at the pattern shown in question **1**. He says, 'There are four hexagons around each square, so the number of hexagons in a pattern must be four times the number of squares.'

Is Sam correct? Give reasons for your answer.

4 Jim and Kim each mix some grey paint using black and white powder paint.

▶ Jim uses 19 grams of white and 6 grams of black.

▶ Kim uses 27 grams of white and 8 grams of black.

Who will make the darker shade of grey? Explain your reasons.

Dimensions and scales

This unit will show you how to:

▶▶ Know and use geometric properties of shapes made from cuboids.

▶▶ Use plans and elevations.

▶▶ Know and use the formula for the volume of a cuboid.

▶▶ Calculate volumes and surface areas of cuboids and shapes made from cuboids.

▶▶ Make simple scale drawings.

▶▶ Reduce a ratio to its simplest form including a ratio expressed in different units.

▶▶ Use units of measurement to estimate, calculate and solve problems.

▶▶ Plot coordinates in all four quadrants.

▶▶ Given the coordinates of points A and B, find the midpoint of the line segment AB.

▶▶ Use bearings to specify direction.

▶▶ Find simple loci, by reasoning, to produce paths and shapes.

▶▶ Use straight edge and compasses to construct:

  ▶ a triangle given three sides (SSS)

  ▶ the perpendicular bisector of a line segment

  ▶ the bisector of an angle.

▶▶ Solve more demanding problems and investigate in the context of shape and space.

▶▶ Represent problems and interpret solutions in graphical form, using correct notation and appropriate diagrams.

▶▶ Use logical argument to establish the truth of a statement.

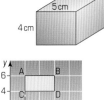

How long now, guv?

We should be able to see land in 5 minutes.

Aeroplanes use angles and distances to calculate directions.

Before you start

You should know how to ...

1 Find the area of a rectangle.

2 Draw a net of a cuboid.

3 Read and plot coordinates.

Check in

1 Find the area of this rectangle:

4.2 m

1.7 m

2 Sketch the net of this cuboid:

8 cm

5 cm

4 cm

3 Write down the coordinates of A, B, C, D.

This spread will show you how to:

▶▶ Know and use geometric properties of shapes made from cuboids.

KEYWORDS

Cube   Vertices
Cuboid   Face
Edge   Isometric
Vertex

A cube has three dimensions: length, width and height.
All its edges are the same length.

It has:

▶ 12 edges   The blue edges are parallel.

▶ 8 vertices   The red edges are perpendicular.

▶ 6 faces   The green edges do not intersect.

You can make other shapes from cubes:

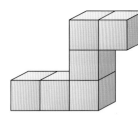

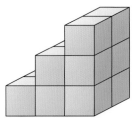

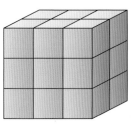

This shape is irregular – there is no pattern to the cubes.

This shape is a prism – it is the same shape throughout the length.

This shape is a cuboid – the layers are rectangular and each has the same number of cubes.

You can draw a cuboid
... on normal paper   or   ... on isometric paper

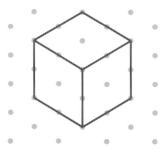

Draw the faces behind one another.
Then join corresponding vertices.

Draw vertical lines vertically and horizontal lines at an angle.

Exercise S4.1

1 You need six multilink cubes.
Make as many different arrangements of six cubes as you can.

These arrangements are the same:

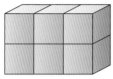

For each of your different arrangements:

▶ Make a sketch on isometric paper.

▶ Classify the arrangement as:
cuboid, prism or irregular.

▶ State the number of edges, vertices and faces.

2 Work with a partner.

▶ Choose one of your shapes from question **1** and, without showing it, describe it to your partner.

▶ They must make the shape from your description.

▶ If the shape is correct, you score one point.

▶ If it is incorrect, your partner scores one point.

Take it in turns to describe your shapes until one of you reaches 8 points.

3 A cube has six different-coloured faces. You can only see three faces at any one time.

Which colours are opposite each other?

4 For any cube, list the number of pairs of edges that:

a are parallel
b are perpendicular
c do not meet.

5 This cube has opposite faces the same colour.

a How many edges are there where the same coloured faces meet?

b How many edges are there where different coloured faces meet?

You may find it helpful to draw a net of the cube.

6 Use multilink cubes to make a shape that would fit exactly through a hole like this:

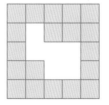

How many different shapes can you make to fit through the hole using exactly 24 cubes?

This spread will show you how to:
▶▶ Begin to use plans and elevations.

KEYWORDS
Plan view   View
Elevation

In this shape made from cubes

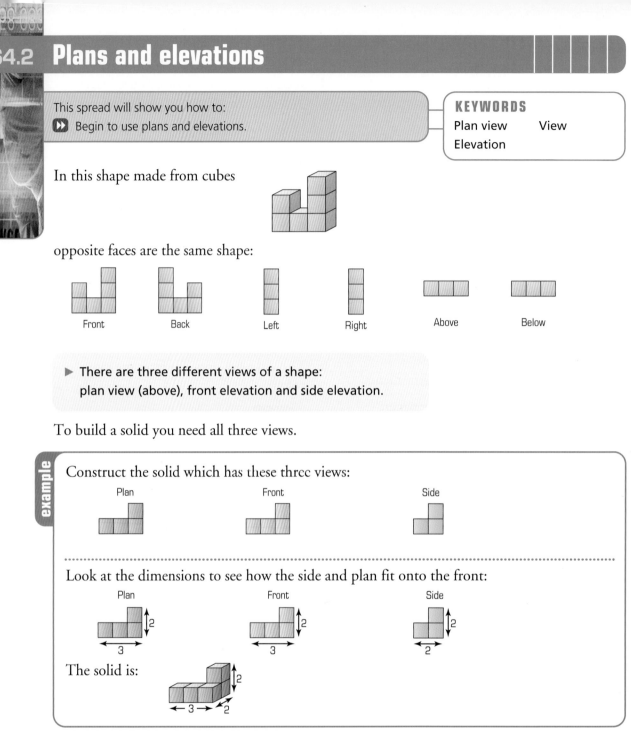

opposite faces are the same shape:

Front   Back   Left   Right   Above   Below

▶ There are three different views of a shape:
plan view (above), front elevation and side elevation.

To build a solid you need all three views.

example

Construct the solid which has these three views:

Plan   Front   Side

Look at the dimensions to see how the side and plan fit onto the front:

Plan   Front   Side

The solid is:

You can make a solid from the plan view so long as you know the heights of each cube in the plan.

This plan ...

3	1	3
	2	1

gives these elevations and ...

Front   Side

makes this solid:

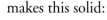

Exercise S4.2

1 For each of these shapes, draw a front elevation, a side elevation and a plan view.

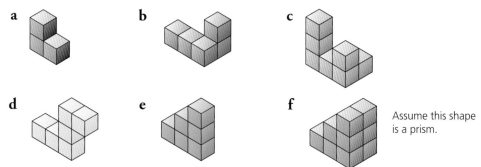

a   **b**   **c**

d   **e**   **f**   Assume this shape is a prism.

2 These diagrams show plan views. For each diagram:
  i Describe what the solid could be.
  ii Give a reason for your answer.
  iii Sketch the 3-D shape if possible.

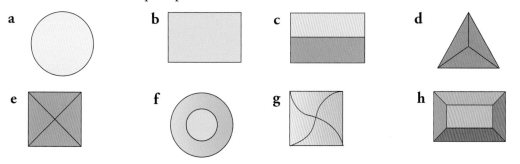

a   **b**   **c**   **d**

e   **f**   **g**   **h**

3 This diagram shows a plan view and the heights of each of the columns.

3	1
1	2
2	

  a Draw the front and side elevations.
  b Construct the shape with multilink cubes and draw it on isometric paper.

4 Here are the three views of two shapes.
Make each shape using multilink cubes and draw it on isometric paper.

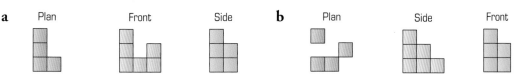

a   Plan   Front   Side   **b**   Plan   Side   Front

Volume and surface area

This spread will show you how to:

▶▶ Know and use the formula for the volume of a cuboid.

▶▶ Calculate volumes and surface areas of cuboids and shapes made from cuboids.

KEYWORDS

Cuboid   Volume

Prism   Surface area

Triangular prism

A cuboid is made up of cubes.
It has a rectangular base and the same number of cubes in each layer:

You can work out the volume by counting the cubes.
A quicker and easier way is to multiply:

▶ **Volume of a cuboid = length × width × height**

Volume is measured in cubic units: cm^3, m^3.
The ^{3} shows there are three dimensions.

The surface area of the cuboid is the sum of the area of all its faces.
The faces are in identical pairs so you only need to work out three areas!

1 Area of plan = length × width

2 Area of front = length × height

3 Area of side = width × height

Area is measured in square units: cm^2, m^2.
The ^{2} shows there are two dimensions.

▶ **Surface area = 2 × plan + 2 × front + 2 × side**

Find the volume and surface area of this prism:

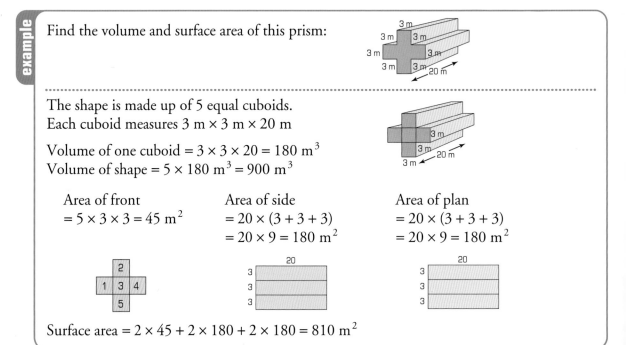

The shape is made up of 5 equal cuboids.
Each cuboid measures 3 m × 3 m × 20 m

Volume of one cuboid = $3 \times 3 \times 20 = 180$ m^3
Volume of shape = 5×180 m^3 = 900 m^3

Area of front
= $5 \times 3 \times 3 = 45$ m^2

Area of side
= $20 \times (3 + 3 + 3)$
= $20 \times 9 = 180$ m^2

Area of plan
= $20 \times (3 + 3 + 3)$
= $20 \times 9 = 180$ m^2

Surface area = $2 \times 45 + 2 \times 180 + 2 \times 180 = 810$ m^2

Exercise S4.3

1 Find the volumes and surface areas of these cuboids:

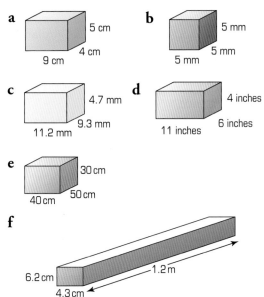

a 5 cm, 4 cm, 9 cm

b 5 mm, 5 mm, 5 mm

c 4.7 mm, 9.3 mm, 11.2 mm

d 4 inches, 6 inches, 11 inches

e 30 cm, 50 cm, 40 cm

f 6.2 cm, 4.3 cm, 1.2 m

2 Find the volume of these prisms:

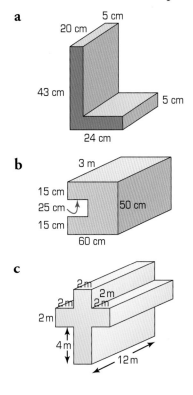

a 5 cm, 20 cm, 43 cm, 5 cm, 24 cm

b 3 m, 15 cm, 25 cm, 15 cm, 50 cm, 60 cm

c 2 m, 2 m, 2 m, 2 m, 2 m, 4 m, 12 m

3 A cuboid has dimensions 14.8 cm by 21.2 cm by 9.8 cm.

   a Estimate its volume.
   b Estimate its surface area.

4 **a** Calculate the surface area of this cuboid:

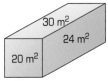

30 m², 24 m², 20 m²

   b Write down the dimensions of the cuboid.
   c Find the volume of the cuboid.

5 **Challenge**
   A cube has a volume of 620 cm³.
   Calculate its surface area.

6 The surface area of a cube is 42 cm².
   Calculate its volume.

7 Calculate the surface area of this triangular prism.

15 m²

8 **a** Calculate the shaded area of the prism.

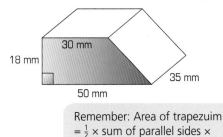

30 mm, 18 mm, 35 mm, 50 mm

> Remember: Area of trapezuim
> = ½ × sum of parallel sides ×
> distance between them

   b Calculate the volume of the prism.

S4.4 Scale drawings

This spread will show you how to:
▶▶ Make simple scale drawings.
▶▶ Reduce a ratio to its simplest form.
▶▶ Use units of measurement to solve problems.

KEYWORDS
Scale drawing
Ratio   Calculate
Estimate

You can estimate an unknown distance using a known distance.

Sally knows it takes her 25 minutes to walk one kilometre.

Joe knows that Emily is 1.5 m tall.

She's been walking for an hour so she's walked about $2\frac{1}{4}$ kilometres.

He estimates that the height of the building is 10 times that, or 15 metres.

▶ **You can draw real-life distances using scale drawings.**

Sally draws a map of her walk.
She uses a scale of 2 cm for 1 km.

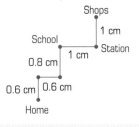

Joe makes a scale drawing of the building.
He uses a scale of 1 cm for 5 m.

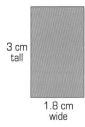

3 cm
tall

1.8 cm
wide

▶ **You can use the scale to find real-life distances from the scale drawing.**

On the map:
From home to the station is 3 cm.

In real life:
2 cm represents 1 km
1 cm represents 0.5 km
3 cm represents 3 × 0.5 km
The distance is 1.5 km.

On the drawing:
The width of the building is 1.8 cm.

In real life:
1 cm represents 5 m
1.8 cm represents 5 × 1.8 m
The distance is 9 m.

You can write the ratios in their simplest form by using the same measures:

  2 cm : 1 km
= 2 cm : 1000 m
= 2 cm : 100 000 cm
  = 1 : 50 000

  1 cm : 5 m
= 1 cm : 500 cm
  = 1 : 500

Exercise S4.4

1 Estimate:
   a the height of your classroom
   b the length of your classroom
   c the distance from school to the nearest shop.
Describe your method of estimating clearly.

2 Choose a suitable scale and use it to make a scale drawing of:
   a your table
   b the classroom whiteboard or blackboard
   c the classroom floor.

3 Write these scales as ratios in their simplest form:
   a a map of a town using 4 cm to represent 1 km
   b a plan of your school grounds using 2 cm to represent 100 m
   c a map of the European Union using 2 cm to represent 100 km.

4 This is a map of Sheffield:

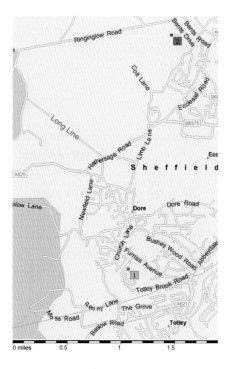

   a Use the scale to calculate the distance 'as the crow flies' between 1 and 2.

   b Estimate the length of Long Line.

   c How long would it take you to run along Long Line?

   d Paul takes four minutes to run one way along Long Line, but eight minutes to run back. Suggest a reason why?

5 Challenge
Kim is lost. She phones Martin for directions to his house. His instructions are:

▶ 'Look straight ahead. Walk until you reach the traffic lights about 250 m away.
▶ Turn left at the lights. Take the third turning on the right, about 750 m up the road.
▶ The road bends to the left and then straightens.
▶ After about 100 m, at the end of the bend there is an alleyway.
▶ About 20 m along the alleyway is a gate. That's my house.'

Make a scale drawing of Kim's journey to Martin's house. Use a sensible scale.

This spread will show you how to:

▶▶ Use units of measurement to estimate, calculate and solve problems.

▶▶ Use bearings to specify direction.

KEYWORDS

Three-figure bearing

Estimate

Calculate

You can get from A to B by:

moving horizontally then vertically ...

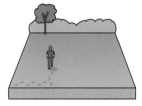

John walks 4 m across the field then 3 m up the field.

or moving directly at an angle.

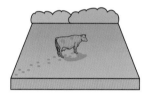

The cow walks 5 m at an angle of 53°.

To specify each path or journey you need to have a reference system.

The simplest one is to use compass directions.

John walks 4 m due East then 3 m due North.

The cow walks 5 m at an angle of 53° clockwise measured from North.

▶ A clockwise angle measured from North is a bearing.

The cow walks 5 m on a bearing of 53°.

It is normal to use three figures for a bearing: 53° is written 053°. You say 'zero five three degrees'.

Bearings are often used to map positions of ships at sea.

example

A ship travels on a three-figure bearing of 060° for 80 kilometres and then on a bearing of 320° for 120 kilometres.

Use a scale drawing to calculate the distance and the bearing from the starting point.

Using a scale of 1 cm for 40 km.

Sketch the journey first.

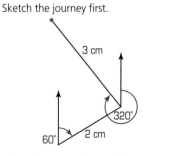

Mark in North. Draw the first stage.

Draw the second stage. Use 360° − 320° = 40°

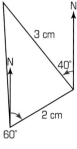

The distance from the start is 3.3 cm which is 3.3 × 40 km.

The angle measured from North is 3° anticlockwise.

The bearing is 357°.

Exercise S4.5

1 The following diagram is drawn using a scale of 1 cm to represent 2 km.

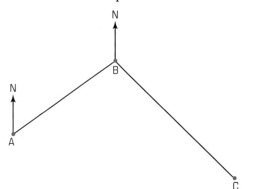

Find:
a the bearing of B from A
b the bearing of C from B
c the distance AB
d the distance AC
e the bearing of C from A
f the bearing of B from C
g the bearing of A from B.

2 The following diagram is drawn using a scale of 1 cm to represent 2 km.

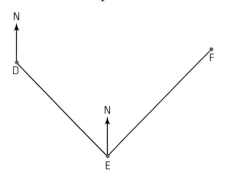

Find:
a the bearing of E from D
b the bearing of D from E
c the bearing of F from E
d the bearing of E from F
e the bearing of F from D
f the distance DE
g the distance EF
h the distance DF.

3 The bearing of A from B is 050°. What is the bearing of B from A?

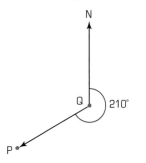

4 The bearing of P from Q is 210°.

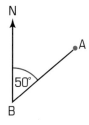

Find the bearing of Q from P.

5 The bearing of X from Y is 140°. Find the bearing of Y from X.

6 Draw a scale diagram using 1 cm to represent 2 km to show this journey:

A speedboat travels:
▶ on a bearing of 130° for 14 km,
▶ then a bearing of 220° for 10 km,
▶ then a bearing of 300° for 12 km.

7 From your answers to question 6, estimate:

a the speedboat's final distance from the starting point
b the speedboat's final bearing from the starting point.

This spread will show you how to:
- ▶▶ Read and plot coordinates in all four quadrants.
- ▶▶ Given the coordinates of points A and B, find the midpoint of the line segment AB.

KEYWORDS
Quadrant
Vertex
Midpoint

You can specify a position or a direction on a coordinate grid:

A = (3, 3) B = ($^-$2, 3)
C = ($^-$4, $^-$1) D = (1, $^-$1)

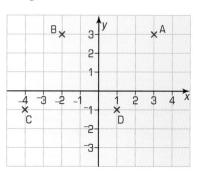

▶ To get from A to B,
move 5 to the left.

▶ To get from B to C,
move 2 to the left and 4 down.

You can join the points to make a
parallelogram ABCD.

The midpoint of a side is halfway along it:

To get from A to B,
move 5 to the left.

To get from A to the midpoint
move $\frac{5}{2}$ units to the left.

▶ The midpoint of AB is (0.5, 3).

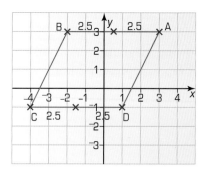

Similarly, the midpoint of CD is ($^-$1.5, $^-$1).

BC is a sloping line.

To get from B to C,
move 2 to the left and 4 down.

To get from B to the midpoint,
move $\frac{2}{2}$ = 1 to the left and $\frac{4}{2}$ = 2 down.

▶ The midpoint of BC is ($^-$2 $-$ 1, 3 $-$ 2) = ($^-$3, 1).

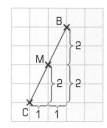

To get from A to D, move 2 to the left and 4
down.
To get from A to the midpoint, move $\frac{2}{2}$ to
the left and $\frac{4}{2}$ down, so move 1 to the left and
2 down.
The midpoint of AD is (2, 1).

Exercise S4.6

1 A and B are two vertices of a triangle. What could the third vertex be if the triangle is:
a right-angled
b isosceles?

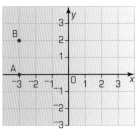

What could the other two vertices be if the quadrilateral is:
c a square
d a rectangle
e a trapezium?

2 Use all four quadrants to draw a shape like this:

Write down the coordinates you have used.

3 The grid shows a line segment AB:

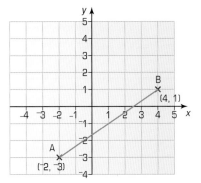

Copy and complete:
To get from A to B, move __ units right and __ units up.
To get half way from A to B, move __ units right and __ units up.
The midpoint of AB is (__, __).

4 Find the midpoints of the lines joining:
a A (6, 3)   B (6, 5)
b C (4, 4)   D (4, 8)
c E (5, 7)   F (9, 7)
d G (6, 3)   H (6, 6)
e I (4, 11)   J (9, 11)
f K (3, 6)   L (7, 10)
g M (4, 7)   N (10, 17)
h P (7, 11)   Q (12, 15)
The first four points are drawn for you:

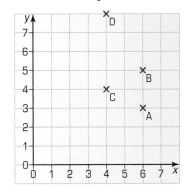

5 Prove that the midpoint of the line segment joining A (x_1, y_1) to B (x_2, y_2) is given by:

$$\left(\frac{x_1 + x_2}{2}, \frac{y_1 + y_2}{2} \right)$$

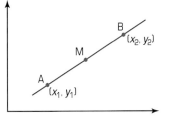

6 A is the point (3, 6) and B is the point (9, 18).
a Find the coordinates of the midpoint M.
b Find the equation of the line joining A and B.

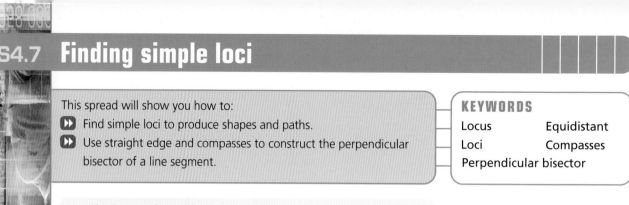

This spread will show you how to:

▶▶ Find simple loci to produce shapes and paths.

▶▶ Use straight edge and compasses to construct the perpendicular bisector of a line segment.

KEYWORDS

Locus   Equidistant

Loci   Compasses

Perpendicular bisector

▶ **The locus of an object is its path.**

The locus of the ball is

a curve.

The locus of the car is

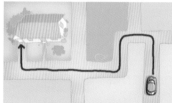

its route.

The locus of the fly is

all over the place!

You can find the locus of points that follow a rule.

example

Mystee is superstitious. When she is in her garden she must always walk exactly the same distance away from the two oak trees.

Show her only path through the garden.

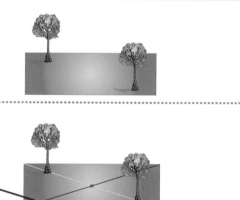

One point that is always equidistant from the two trees is the point exactly in between them.

The path she can take looks like this:

It is the perpendicular bisector of the line joining the two trees.

▶ **The locus of points equidistant from two fixed points is the perpendicular bisector of the line joining the fixed points.**

You use compasses to construct a perpendicular bisector:

Draw an arc from one point ...

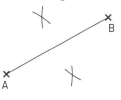

... and an equal arc from the other point.

Join the intersections.

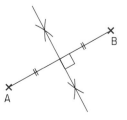

Exercise S4.7

1 **a** Describe your journey to school as a precise route.
  b Now describe your journey from school to home.

2 Describe five everyday examples of loci.

3 You need one red, one blue and several white counters.
  a Place the red and blue counters on a table.

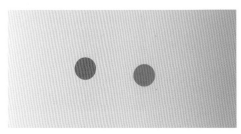

  Now use the white counters to show the locus of all points equidistant from the red and blue counters.

  b Place the red and blue counters like this on your table:

  Use the white counters to show the locus of all points equidistant from the red and blue counters.

  c Investigate different positions of the red and blue counters.
  Explain your findings.

4 Copy the diagram and construct the locus of all points equidistant from A and B.

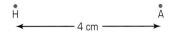

A   8 cm   B

5 In a 'Find the ball' competition, you know that the ball is always the same distance from the two players, Halima and Alison.

  Copy the diagram and construct the locus of the ball.

6 Two trees are 12 m apart.

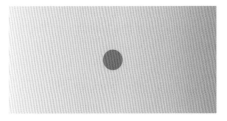

  Draw a scale drawing to show a path which is an equal distance from the two trees.

7 Place a blue counter in the middle of your table.

  Use white counters to show the locus of points at a fixed distance from the blue counter.

S4.8 More loci

This spread will show you how to:
▶▶ Find simple loci to produce paths and shapes.
▶▶ Use straight edge and compasses to construct the bisector of an angle.

KEYWORDS
Locus   Bisector
Loci   Straight edge
Equidistant   Compasses
Adjacent

The locus of points that are equidistant from two fixed objects
... is the perpendicular bisector of the line joining the objects.

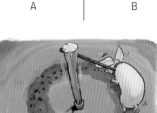

Here are some other common loci.

The distance from one fixed point

Porky the pig eats too much.
He is on a lead that is attached to a post.
He can only move in a circle.

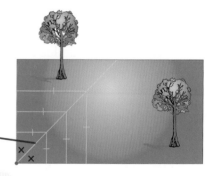

▶ The locus of points that are the same distance from a fixed point is a circle.

Equidistant from two adjacent edges

Mystee's dad, Fred, is also superstitious.

He will only walk across the garden so that he is always the same distance from the left and bottom edges.
His path bisects the angle between them.

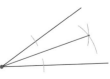

▶ The locus of points that are the same distance from two adjacent edges is the bisector of the angle between the edges.

You construct an angle bisector like this:

Use compasses to draw arcs on each arm.

Draw arcs from these arcs.

Use a straight edge to join the last arcs and the angle.

Exercise S4.8

1 a A donkey is tethered to a post with a 12 m piece of rope.

Using a scale of 1 cm to 2 m draw a diagram to show all the grass the donkey can reach to eat.

b Construct the locus of all points 6 cm from a fixed point.

2 Copy the diagram and construct the locus of all points equidistant from A and B.

A ⟵ 10 cm ⟶ B

3 A radio mast can transmit within a 30 km radius.

Draw a scale drawing to show the region that can receive the radio signal. Remember to give your scale.

4 Copy or trace these diagrams.
Construct the locus of points which are equidistant from the edges of these fields.

a **b**

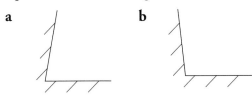

5 a Two competing radio stations, Town FM and City Radio are 20 km apart.
Town FM can transmit within a 30 km radius.
City Radio can transmit within a 20 km radius.

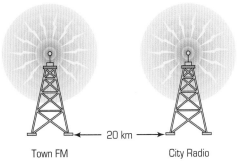

Town FM ⟵ 20 km ⟶ City Radio

Draw the region that can only listen to Town FM.

b Prettygood Pizzas deliver within a 3 km radius.
Super Pizzas deliver within a 4 km radius.
The shops are 5 km apart. Show the region which both pizza shops deliver to.

6 The top scorer for the Didcot Dynamos basketball team can shoot a basket from 16 m.

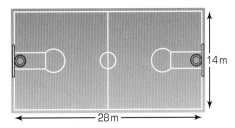

14 m
⟵ 28 m ⟶

Copy the plan of the basketball court and show from what area of the court she can score directly.

Constructing triangles

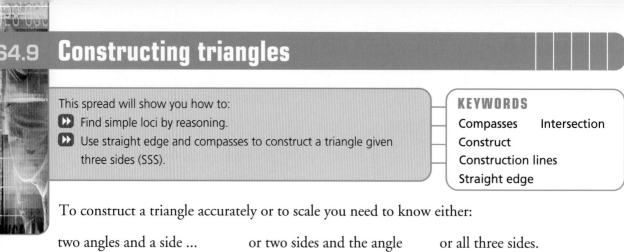

This spread will show you how to:
- ▶▶ Find simple loci by reasoning.
- ▶▶ Use straight edge and compasses to construct a triangle given three sides (SSS).

KEYWORDS

Compasses   Intersection
Construct
Construction lines
Straight edge

To construct a triangle accurately or to scale you need to know either:

two angles and a side ...   or two sides and the angle   or all three sides.
                                 between them ...

When you know two angles you
know all three.

To construct a triangle given three sides you need to use a straight edge and compasses.

example

Construct triangle ABC where AB = 8 cm, BC = 5 cm and AC = 6 cm.

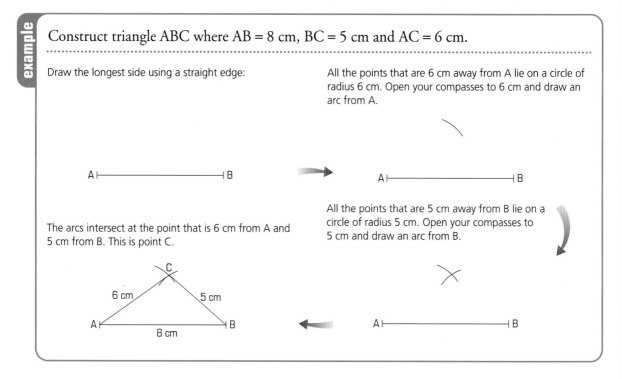

Draw the longest side using a straight edge:

All the points that are 6 cm away from A lie on a circle of radius 6 cm. Open your compasses to 6 cm and draw an arc from A.

The arcs intersect at the point that is 6 cm from A and 5 cm from B. This is point C.

All the points that are 5 cm away from B lie on a circle of radius 5 cm. Open your compasses to 5 cm and draw an arc from B.

Leave your construction lines on your drawing to show your method.

Exercise S4.9

1 Construct these triangles accurately:

a

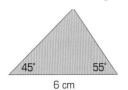

45°   55°
6 cm

b

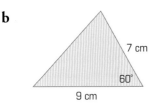

7 cm
60°
9 cm

c

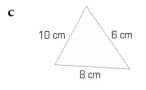

10 cm   6 cm
8 cm

d

9 cm
52°   67°

e

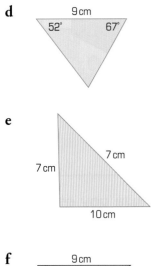

7 cm   7 cm
7 cm
10 cm

f

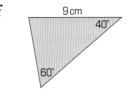

9 cm
40°
60°

2 Construct triangle PQR where
PQ = 5 cm, PR = 6 cm and QR = 7 cm.

Measure the angles of the triangle.

3 **a** Construct triangle ABC where
AB = 4 cm, BC = 6 cm and
AC = 6 cm.
Measure the angles of the triangle.

b What type of triangle is it?

4 This sketch shows the area lit up by
a headlamp.

5 m

22 m   18 m

Draw a scale drawing of this area.
Use a scale of 1 cm = 2 m.

5 Construct this quadrilateral.

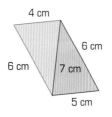

4 cm
6 cm
6 cm   7 cm
5 cm

6 **a** Construct triangle ABC where
AB = 10 cm, AC = 4 cm and
BC = 5 cm.

b What goes wrong?
Explain why it does not work.

7 Can you construct a triangle with sides
12 cm, 6 cm and 6 cm?

Give a reason for your answer.

8 Can you construct a triangle with sides
12 cm, 6 cm and 7 cm?
Give a reason for your answer.

Formulating hypotheses

This spread will help you to:

▶▶ Discuss an issue that can be addressed by statistical methods and identify related questions to explore.

▶▶ Decide which data to collect to answer a question, the degree of accuracy needed, and identify possible sources.

▶▶ Plan how to collect the data, including sample size.

KEYWORDS

Hypothesis   Primary data
Hypotheses   Sample
Secondary data

Artists and engineers study connections between body measurements.

Detectives or scientists might want to estimate somebody's height based on the length of a footprint!

Vicky and Simon are investigating the ratio of height to foot length. They are discussing how to specify the problem.

I think that tall people have big feet.

Well, my brother is tiny and he's got huge feet!

They decide on their hypothesis:

> The connection between foot size and height is good enough to make a prediction of somebody's height from their foot size.

Vicky and Simon decide to start by collecting primary data from their class. This is a good way to get an idea about any connection.

A bigger sample will give more reliable results. They need to find a good source of secondary data to confirm their results, or check whether they apply to other age groups.

The sample size is the number of data values you collect.

Primary data is:
▶ Easy to collect.
▶ Organised the way you want it.

Primary data might:
▶ Give you a small sample size.
▶ Relate to a specific group only – like your own class members.

Secondary data is:
▶ Often harder to find.
▶ Organised the way someone else wanted!

Secondary data might:
▶ Give you a larger sample size.
▶ Relate to wider groups than you have easy access to.

Exercise D3.1

The cards show different topics that can be investigated using statistical methods.
Each card gives some ideas about aspects of the topic that you could investigate.

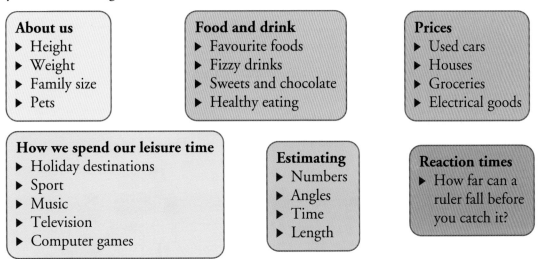

About us
▶ Height
▶ Weight
▶ Family size
▶ Pets

Food and drink
▶ Favourite foods
▶ Fizzy drinks
▶ Sweets and chocolate
▶ Healthy eating

Prices
▶ Used cars
▶ Houses
▶ Groceries
▶ Electrical goods

How we spend our leisure time
▶ Holiday destinations
▶ Sport
▶ Music
▶ Television
▶ Computer games

Estimating
▶ Numbers
▶ Angles
▶ Time
▶ Length

Reaction times
▶ How far can a ruler fall before you catch it?

These questions are designed to help you get started on a statistical project.
You should work through them all in turn.

1 **Choose one issue** from the cards that you could investigate statistically.
You can choose an issue of your own instead!

2 **Specify the problem** that you will investigate. Decide on a hypothesis that you will test, and write it down as clearly as possible.

3 **Decide what data you will need to collect** to test your hypothesis. Write a description of the data you will need, and where you will get it from. Make it clear whether you will be using primary data or secondary data, and explain how big your data sample will be.

4 **Write an explanation of the decisions you made** in questions **1–3**.
This will form an introduction to your project.

> If you have time, you could choose several different areas to investigate, and work through each question. You could then select the option that looks most promising to develop further!

This spread will help you to:
▶▶ Plan how to collect the data, including sample size.
▶▶ Construct frequency tables with equal class intervals.
▶▶ Collect data using a suitable method.
▶▶ Construct frequency diagrams for continuous data.

KEYWORDS

Sample   Data log
Frequency table
Questionnaire

You need to decide on the best way to collect any primary data for your project.
You may need to compile a questionnaire.

Other ways of collecting primary data include:

Observation

You just sit and watch something happen, and record the results.

Controlled experiment

You set up a situation to produce the data needed. You could investigate the relationship between load and extension for a spring.

Data logging

Data is collected automatically by instruments and can be sent to a computer.

Some useful sources of secondary data are newspapers, magazines, reference books and the internet.

Simon and Vicky collected the data they needed from their class.

For each member of their class, they measured height and foot length.
They decided how precise their measurements needed to be.

▶ For height, they measured to the nearest centimetre.
▶ For foot length, they measured to the nearest millimetre.

They used a data collection sheet for the paired data:

Name	Foot length (cm)	Height (cm)	Name	Foot length (cm)	Height (cm)	Name	Foot length (cm)	Height (cm)
Abby	22.6	143	Jez	27.0	167	Rita	22.4	157
Ben A	24.0	170	Karla	24.7	154	Steven	23.2	162
Ben J	22.3	115	Leo	22.1	155	Tom	24.6	154
Barry	19.8	150	Liam	20.0	158	Tyrone	24.0	132
Ellie	24.1	169	Mina	25.5	171	Uma	22.7	145
Fatima	22.9	154	Nate	21.6	169	Vera	23.2	153
Harry	20.9	150	Oscar	28.0	172	Will	25.4	158
Ina	23.8	157	Pete B	24.1	161	Yvonne	22.3	138
Iona	25.0	170	Pete S	21.3	142	Zoe	24.9	149

Exercise D3.2

1 Before you collect data for your own project, you will need to design a good questionnaire or data collection sheet.

▶ Decide which type of data collection your project needs:

observation, a controlled experiment, data logging or a questionnaire.

▶ List all the items of data you will need to collect.
▶ Decide on the degree of accuracy each of your measurements will need.

Now design a draft version of your questionnaire or data collection sheet.

2 Carry out a pilot data collection:

Try out your data collection sheet or questionnaire with a small sample size – ask a few people or record a few measurements.
Here are a few suggestions for the sort of things to look for:

▶ How easy was it to complete each part of your data collection?

▶ For questionnaires:
Were people able to answer each question clearly?
Do any of your questions need to be narrowed down to a range of options?

▶ For data collection sheets:
Have you decided on a suitable degree of accuracy for any measurements you are making?
Was there enough space on the sheet to record the data?

▶ Have you forgotten anything?
Think carefully about whether there is any other data you need to collect, and add any extra questions you need.

▶ You could construct a simple frequency diagram to see if you have collected all the information you need to draw the diagram.

3 Make a final version of your questionnaire or data collection sheet and carry out your data collection.

A pilot data collection is a 'try-out' of the collecting method with a small sample size.
It will show you if there are any major problems with your survey.

This spread will help you to:
▶▶ Calculate statistics, including with a calculator.
▶▶ Calculate a mean, using an assumed mean.
▶▶ Know when to use the modal class for grouped data.
▶▶ Compare two distributions using the range and one or more of the mode, median and mean.

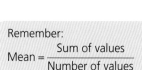

KEYWORDS

Mean	Median
Modal class	Mode
Range	
Assumed mean	

Vicky and Simon want to find the average height of people in their survey.
There are three averages they can use: mean, median and mode.

When the values are large, it helps to use an assumed mean:
▶ assume a convenient value for the mean
▶ subtract it from each piece of data
▶ find the mean of the new values
▶ add the assumed mean back on.

Remember:
$$\text{Mean} = \frac{\text{Sum of values}}{\text{Number of values}}$$

example

Find the mean height from Vicky and Simon's data:

143	170	115	150	158	169	154	150	157	170	167	154	155	158
171	169	172	161	142	157	162	154	132	145	153	158	138	149

Assume a mean of 150 cm. Subtract 150 from each value:

$^-7$	$^+20$	$^-35$	0	$^+8$	$^+19$	$^+4$	0	$^+7$	$^+20$	$^+17$	$^+4$	$^+5$	$^+8$
$^+21$	$^+19$	$^+22$	$^+11$	$^-8$	$^+7$	$^+12$	$^+4$	$^-18$	$^-5$	$^+3$	$^+8$	$^-12$	$^-1$

Find the mean of these values:
   The positive differences make $^+219$, and the negative differences make $^-86$.
   The total is $219 - 86 = 133$
   The mean is $133 \div 28 = 4.75$ cm

Add the assumed mean back on: $150 + 4.75 = 154.75$ cm
The mean height is 154.75 cm.

You can use this formula to find the mean using an assumed mean:

$$\text{Mean} = \text{Assumed mean} + \frac{\text{Sum of differences}}{\text{Number of data values}}$$

To find the median you can draw a stem-and-leaf diagram.

It is not appropriate to find the mode as there are lots of different values, but you can put the data into classes, and find the modal class.

Look back to D2.4 to find out how to draw a stem-and-leaf diagram.

Exercise D3.3

1 For the data on page 232, find the median height.
Put the data into classes and find the modal class.
Which average do you think best illustrates the average height?

2 The lengths (in cm) of 10 pieces of wood are recorded in the table.
The second row shows how much longer than 1m each piece is.

Length (cm)	107	111	103	124	100	106	104	104	108	119
Difference from 1 m (cm)	7	11								

  a Copy and complete the table.
  b Work out the total of the differences in the second row.
  c Work out the mean length of the 10 pieces of wood, using 100 cm as an assumed mean.
  d Work out the mean length of the pieces of wood without an assumed mean.
    Which method was easiest? Why?

3 The numbers of pages in each of eight books are:
324, 311, 308, 369, 318, 312, 330, 324.
Work out the mean number of pages per book.
Use an assumed mean of 300 pages.

4 The numbers of words in 10 sentences from a book are:
14, 27, 17, 27, 20, 25, 39, 28, 40, 29.
Copy and complete the table, which shows the differences from
an assumed mean of 25 words per sentence.

Length (cm)	14	27	17	27						
Difference from 1 m (cm)	⁻11	⁺2	⁻8							

Use your table to find the mean number of words per sentence.

5 For each of these two data sets:
  a Calculate the mean using an appropriate assumed mean.
  b Explain how you choose the assumed mean in each case.
  c Compare the two parts of the data sets using the mean and the range.

    i Weight (in kg) of the players in a football team and a rugby team
    Football team   75, 68, 77, 91, 73, 82, 81, 85, 75, 88, 83.
    Rugby team   84, 93, 76, 94, 98, 87, 83, 85, 94, 85, 86, 92, 92, 99, 81.

    ii Length (in m) of copper tubes in two samples
    Sample A   1.98, 2.03, 1.99, 2.05, 2.04, 1.96, 2.00, 2.03, 1.98, 1.99.
    Sample B   1.91, 2.01, 2.04, 1.87, 2.10, 2.14, 2.08, 1.89, 1.93, 1.97.

6 Find the mean, median and mode of the foot length data on page 230.
Which is the best average to use? Why?

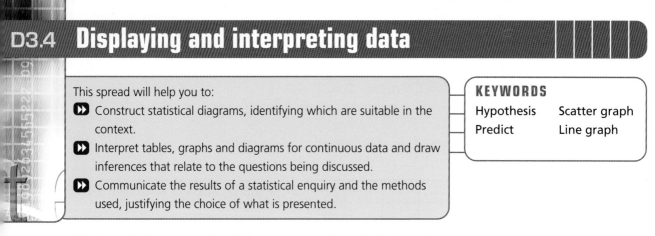

This spread will help you to:

▶▶ Construct statistical diagrams, identifying which are suitable in the context.

▶▶ Interpret tables, graphs and diagrams for continuous data and draw inferences that relate to the questions being discussed.

▶▶ Communicate the results of a statistical enquiry and the methods used, justifying the choice of what is presented.

KEYWORDS

Hypothesis   Scatter graph
Predict   Line graph

Vicky and Simon wonder if the results confirm their hypothesis:

> The connection between foot size and height is good enough to make a reasonable prediction of somebody's height from their foot size.

To show the connection they use a scatter graph:

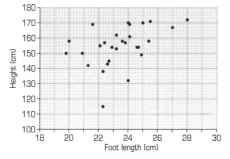

Each dot represents a person in the project.

The distribution of dots on the scatter graph shows there is some connection between foot length and height – but it is not easy to predict from.

They compare their data with some secondary data from the Census at School website:

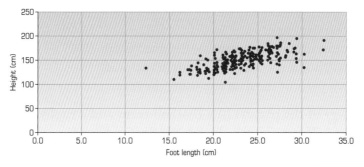

Vicky and Simon decide that the connection is strong enough to be able to make useful predictions.

To experiment, they ask Gemma to collect some more data on foot lengths, and they use the scatter diagram to predict the heights.

They will work out the percentage error of the predictions compared with the actual heights.

Exercise D3.4

Class 8S did project work about transport in the UK.

1 Jessica looked at the cost of second-hand cars.

> Hypothesis: The age of a second-hand car tells you how much you should expect to pay for it.

Data collection
She collected age and price details for a sample of second-hand cars from the internet.

Data display
The scatter diagram shows the data she collected:

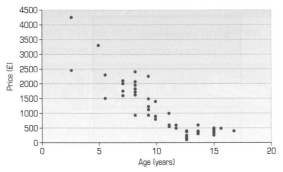

a Write a conclusion for Jessica's project.
  - ▶ You should say whether the data supports her hypothesis.
  - ▶ You could include suggestions for further research.

b Is the graph shown an effective way of showing Jessica's data? What changes would you make, if any?

2 Wesley used this line graph in his project.

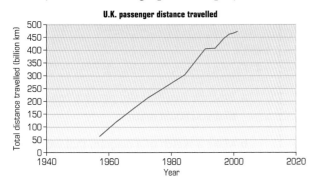

a Suggest a hypothesis for Wesley's project. Write a conclusion based on the hypothesis and the information shown in the graph.

b Suggest other hypotheses for Wesley to test to develop his project. Explain what data would be needed to test each hypothesis.

This spread will help you to:
- ▶▶ Appreciate that random processes are unpredictable.
- ▶▶ Understand that if an experiment is repeated there may be, and generally will be, different outcomes.
- ▶▶ Compare experimental and theoretical probabilities.

KEYWORDS

Experiment	Probability
Model	Simulation

'Double or Bust' is a game show.

A large number of green balls and red balls are placed in the hopper. There are the same number of reds and greens.

The contestant presses the 'Roll' button to release a ball, which rolls down to the end of the chute.

STICK   ROLL

Here are the rules of the game:

The first ball is worth £100.
After each roll the contestant has to choose whether to stick or roll.

- ▶ If you stick, you keep the money and the game is over.
- ▶ If you roll, another ball is released.
- ▶ If it is the same colour, the prize money is doubled.
- ▶ If it is the other colour the prize goes back to £100.

The game ends when the contestant sticks, or when there are 12 balls in the chute.

You can model the game to find the best strategy.
You design an experiment to capture the key features.

You can use:

A coin	A dice	A computer
Spinning the coin represents a roll of a ball. Use heads for red, and tails for green.	Put coloured stickers on the faces, or just use odd for red and even for green.	You can generate a lot of data quickly, but it will be harder to set up.

As there are a large number of balls in the hopper you can assume that the probability of getting each colour is 0.5 or $\frac{1}{2}$ each roll.

The exercise will help you understand what is a good result and when it would be sensible to stick.

Exercise D4.1

1 The company that makes the game show 'Double or Bust' claims that the top prize is over £200 000.
   a Explain how they worked this out.
   b How likely is it that a contestant would win this top prize?

2 Model the game using a coin or a dice.
Play the game three or four times to get a feel for the model.
You do not need to write anything down.

3 Model the game in a systematic way.
The question you really need to answer is:
What is the longest run of the same colour that you would expect, if you can roll up to 12 times?

> If you approach the problem this way, you don't need to worry about deciding to stick or roll.

- Roll 12 times, and record how many runs of various lengths you get.
- Repeat the experiment a large number of times.

The data should help you decide how long a run you would expect to get in 12 rolls.

Copy this data collection sheet:

Trial	Data	Run length											
		1	2	3	4	5	6	7	8	9	10	11	12
1	RRGGRGRRRRGG	2	3		1								
2	GGGRRRGGGRGG	1	1	3									
3	RRRRGRGGRRGG	2	3		1								
4													

There were 3 runs of 2 in this set.   The longest run was 4 reds.

- Replace the sample data with the results of your own experiments.
- Shade the longest runs to make them easy to see at a glance.

4 The data you collected in question **3** should tell you how long a run you can reasonably expect in 12 rolls.
You can use this data to consider when it is sensible to stick.
Play the 'real' game again as in question **2**.

> Once you've got a good run, you can consider sticking!

- After each roll, decide whether to stick or roll.
- Use the data you collected for question **3** to decide on the best strategy.
- Record your results.

What prize money can you reasonably expect to win?

This spread will help you to:

➤➤ Know that if the probability of an event occurring is p, then the probability of it not occurring is $1 - p$.

➤➤ Estimate probabilities from experimental data.

➤➤ Compare experimental and theoretical probabilities in different contexts.

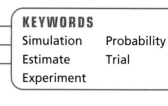

KEYWORDS

Simulation   Probability

Estimate   Trial

Experiment

The more data you collect, the more reliable your results.

A good way of doing this quickly is to use a computer simulation.

The table shows the longest runs in 1000 games of 12 rolls, simulated by a computer program.

Longest Run	1	2	3	4	5	6	7	8	9	10	11	12
Frequency	1	125	356	241	151	68	25	17	10	3	2	1

The bar chart shows that the most likely longest run is 3.

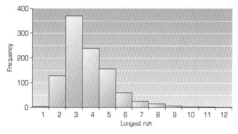

It is possible for the longest run to be 1 or 12, but it is very unlikely.
A run length of 12 happened just once in 1000 simulated games.

You can use the values in the table to estimate probabilities.
Use the formula:

▶ Experimental probability $= \dfrac{\text{Number of successful trials}}{\text{Total number of trials}}$

example

Estimate the probability of the longest run being:

a Exactly 3   **b** 4 or more

....

a The number of runs of 3 was 356.
The probability of a run of 3 is $\frac{356}{1000} = 0.356$

b You can work out:
▶ the number of runs that were 4 or more
▶ or the number of runs that were 3 or less and then subtract from the total:
$1 + 125 + 356 = 482$ so probability of 3 or less $= 0.482$.
Probability of 4 or more $= 1 - 0.482 = 0.518$ – more than an even chance.

This suggests that a good strategy might be to stick after a run of 4.

Exercise D4.2

1 This table shows the results of 30 longest run experiments:

Longest run	2	3	4	5	6	7	8	9
Frequency	5	10	7	4	2	1	0	1
Probability	0.17							

The experimental probability of the longest run being exactly 2 is

$\frac{5}{30} = 0.17$ (2 dp).

a Copy and complete a table like this to show the experimental probabilities based on the data you collected in D4.1 question **3**.

b Use the table to estimate the probability of getting each longest run.

2 Collect together as much longest run data as you can from other students in your class.

a Make a new table, like the one in question **1**, to show the experimental probability for each longest run, based on the data from the whole of your class.

b Draw a bar chart to show the frequencies of different longest runs.

3 This table shows the results of a computer simulation of 5000 games, which recorded the longest run in each game.

Longest run	1	2	3	4	5	6	7	8	9	10	11	12
Frequency	2	552	1722	1350	788	327	149	68	22	11	2	7
Probability												

a Copy the table, and fill in the bottom row to show the estimated probability of each longest run.

b Use your table to work out the experimental probability that the longest run will be 4 or more.

c Draw a bar chart to show the frequency data from the table.

4 In questions **1**, **2** and **3**, you worked out three sets of experimental probabilities, based on three different sets of data. Which set of estimates should be most reliable? Explain your answer.

5 Explain how your strategy in the game would change if the rules were different.

What would you do if:

a You could roll more balls – for example, 20 instead of 12?

b The prize money was tripled for each ball of the same colour, instead of doubled?

This spread will help you to:
▶▶ Communicate orally and on paper the results of a statistical enquiry.
▶▶ Estimate probabilities from experimental data.
▶▶ Understand that increasing the number of times an experiment is repeated generally leads to better estimates of probability.

KEYWORDS
Experiment   Probability
Bias   Simulation

Monty Hall is a game show host.

He gives a contestant a choice of 3 doors.

The star prize (a luxury car) is hidden behind one of the doors.

There is a goat behind each of the others!

Monty knows what is behind each door.

After the contestant has made their choice, Monty opens one of the other doors to reveal a goat.

There are now just two doors to choose from.

The contestant is given the choice of sticking with their original door, or swapping to the other one.

People have different views about whether to stick or swap:

I would always stick. There's no point changing, because you know from the start that Monty will show you a goat!

It doesn't matter whether you stick or switch – with two doors left, you've got a probability of 50% of winning.

I would swap – after all, you might get lucky!

You can model the situation, and carry out an experiment to help you to decide the best strategy.

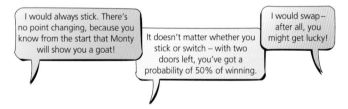

▶ Find out what happens if you stick with your original choice.
▶ Find out what happens if you swap.
▶ Estimate the probability of winning the car with each strategy.

The exercise will help you to design the experiment carefully, to make sure that it is fair, and avoids bias.

Exercise D4.3

1 Here is one way to simulate the start of the experiment.

> ▶ Monty rolls a dice to decide which door to put the car behind.
> ▶ The contestant rolls a dice to decide which door to pick.

  a Explain why it is important to use apparatus to choose the doors.
  b Explain how to use a dice to choose the doors.

2 Monty opens one of the doors.
He will not choose the one the contestant picked or the one that the car is behind.
Explain which door or doors Monty could open when:
  a The car is behind door 2. The contestant chose door 3.
  b The car is behind door 1. The contestant chose door 2.
  c The car is behind door 3. The contestant chose door 3.

3 Explain why this design for the experiment would not be very good:
Monty writes the number of the winning door on a piece of paper.
The contestant writes down a guess on another piece of paper.
If the guess is correct, Monty tosses a coin to decide which door to open.
If the guess is wrong, he opens the other 'wrong door'.

4 Work in small groups for this activity.
One person will play the role of Monty, and one person will be the contestant.
You will need to agree how to carry out the experiment.

Remember:

> ▶ Monty must choose the door that the star prize is behind, and not change it.
> ▶ The contestant must not know where the prize actually is until the final choice is made.

> Decide on your strategy before you start.
> You could carry out a series of trials with an 'always stick' strategy, and another series with an 'always swap' strategy.

Carry out some trials and record the results.
You will need a table like this to record your results:

Strategy: Always stick with original choice					
Trial	Correct door	First choice	Door opened	Final choice	Win or lose
1	1	1	3	1	Win
2	3	2	1	2	Lose
3					

5 Once you have collected as much data as possible, work out an estimate for the probability of winning the car with each strategy. Explain which strategy you would advise a contestant to follow.

Glossary

area: square millimetre, square centimetre, square metre, square kilometre
S2.1, A5.1

The area of a surface is a measure of its size.

arithmetic sequence
NA1.6

In an arithmetic sequence each term is a constant amount more or less than the previous term.

ascending, descending
A5.4

Ascending means going up or getting bigger. Descending means going down or getting smaller.

associative
NA1.1

Addition and multiplication are associative because it doesn't matter which order you add or multiply the numbers in.
$(1 + 2) + 3 = 1 + (2 + 3)$
$(1 \times 2) \times 3 = 1 \times (2 \times 3)$.

assumed mean
D3.3

An assumed mean is used to simplify the arithmetic when calculating the mean. The assumed mean is subtracted from all the data and added back on once the mean of the smaller numbers has been calculated.

average
D2.5

An average is a representative value of a set of data.

axis of symmetry
S3.2, S3.4

An axis of symmetry of a shape is a line about which the shape can be folded so that one half fits exactly on top of the other half.

axis, axes
A3.4, A5.7

An axis is one of the lines used to locate a point in a coordinate system.

bar chart
D2.1

A bar chart is a diagram that uses rectangles of equal width to display data. The frequency is given by the height of the rectangle.

bar-line graph
D2.1

A bar-line graph is a diagram that uses lines to display data. The lengths of the lines are proportional to the frequencies.

base (of plane shape or solid)
S4.2, S4.3

The lower horizontal edge of a plane shape is usually called the base. Similarly, the base of a solid is its lower face.

base

bearing, three-figure bearing
S4.5

A bearing is measured from the North in a clockwise direction.
The bearing of B from A is 045°.

N
B
45
A

best estimate
S2

The best estimate of a value is the closest you can achieve.

bias
D4.3

An experiment or selection is biased if not all outcomes are equally likely.

billion
N3.1

One thousand million or 1 000 000 000.

bisect, bisector
S1.4, S1.5, S4.8

A bisector is a line that divides an angle or another line in half.

brackets
N4.3, A5.3

Operations within brackets should be carried out first.

calculate, calculation
N4.5, S4.4, S4.5

Calculate means work out using a mathematical procedure.

calculator: clear, display, enter, key, memory
NA1.4

You can use a calculator to perform calculations.

cancel, cancellation
N2.3

A fraction is cancelled down by dividing the numerator and denominator by a common factor.

For example, $\frac{24}{40} \overset{\div 8}{\underset{\div 8}{=}} \frac{3}{5}$

capacity: litre
S2

Capacity is a measure of the amount of liquid a 3-D shape will hold.

centre of enlargement
S3.6

The centre of enlargement is the point from which an enlargement is measured.

Centre of enlargement

centre of rotation
S3.2

The centre of rotation is the fixed point about which a rotation takes place.

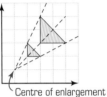

certain
D1

An event that is certain will definitely happen.

chance
D1

Chance is the probability of something happening.

class interval
D3.3

A class interval is a group that you put data into to make it easier to handle.

collect like terms
A2.3, A2.4, A4.2, A5.1,

Collecting like terms means collecting all similar terms together. For example, $2x + 3x + 4 + 2 = 5x + 6$.

common factor
NA1.3

A common factor is a factor of two or more numbers. For example, 2 is a common factor of 4 and 10.

commutative
N4.1, N4.2

An operation is commutative if the order of combining two objects does not matter. For example, addition is commutative as $4 + 3 = 3 + 4$, but subtraction is not commutative because $4 - 3 \neq 3 - 4$.

compare
N2.6, N3.7

Compare means to assess the similarity of.

compasses (pair of)
S1.5, S1.6, S4.7, S4.8, S4.9

Compasses are used for constructions and drawing circles.

discrete (data)
D2.2

Discrete data can only take certain definite values, for example, integers between 10 and 20.

displacement
S3.2

Displacement is a measure of how something has been moved.

distance
S4.5

The distance between two points is the length of the line that joins them.

distance–time graph
A3.5

A graph showing distance on the vertical axis and time on the horizontal axis.

distribution
D2.5

Distribution describes the way data is spread out.

distributive
N3.8

Multiplication is distributive over addition and subtraction. For example, $a(b + c) = ab + ac$.

divide, division
NA1.2, N3.4, N3.6, N4.2, N4.5

Divide ($\div$) means share equally.

divisible, divisibility
N3.6

A whole number is divisible by another if there is no remainder left.

divisor
N3.6

The divisor is the number that does the dividing. For example, in $14 \div 2 = 7$ the divisor is 2.

double, halve
NA1.5

Double means multiply by two. Halve means divide by two.

draw
S4.2, S4.4

Draw means create a picture or diagram.

edge (of solid)
S4.1

An edge is a line along which two faces of a solid meet.

elevation
S4.2

An elevation is an acurate drawing of the side or front of a solid.

enlarge, enlargement
S3.5

An enlargement is a transformation that multiplies all the sides of a shape by the same scale factor.

equal (sides, angles)
S3.1

Equal sides are the same length. Equal angles are the same size.

edge

equally likely
D1.1, D1.3

Events are equally likely if they have the same probability.

equals (=)

Equals means having exactly the same value or size.

equation
A2.1, A3.4, A4.1, A5.2, A5.3
A5.4, A5.8

An equation is a statement linking two expressions that have the same value.

equation (of a graph)
A5.5

The equation of a graph links the two variables together to give co-ordinates.

equidistant
S4.7, S4.8

Equidistant means the same distance apart

equivalent, equivalence
N2.2, N2.4, A2.1, S2.2, N3.5
N4.1, N4.4, N4.5, A5.1, A5.2

Equivalent fractions are fractions with the same value.

estimate
D1.4, N3.3, N3.5, N3.6, N4.4
N4.5, S4.4, S4.5, D4.2

An estimate is an approximate answer.

evaluate

Evaluate means find the value of an expression.

event
D1.1, D1.2

An event is an activity or the result of an activity.

exact, exactly
N3.2

Exact means completely accurate.
For example, three divides into six exactly.

exceptional case
P1.6

An exceptional case is one which is used to disprove a rule or
hypothesis.

experiment
D1.5, D1.6, D4.1, D4.2, D4.3

An experiment is a test or investigation to gather evidence for or
against a theory.

experimental probability
D1.4, D1.5, D1.6

Experimental probability is calculated on the basis of the results of
an experiment.

expression
A2.1, A2.5, A5.1, A5.2

An expression is a collection of numbers and symbols linked by
operations but not including an equals sign.

exterior angle
S1.2

An exterior angle is made by extending one side of a shape.

face
S4.1

A face is a flat surface of a solid.

face

factor
NA1.3, N3.4

A factor is a number that divides exactly into another number.
For example, 3 and 7 are factors of 21.

factorise
A5.1

A number or expression is factorised when it is written as a product
of its factors.

fair, biased
D1.6, D4.3

In a fair experiment there is no bias towards any particular outcome.

favourable outcome
D1.3

A favourable outcome is a successful result of doing something. For
example, throwing a 'six' with a fair dice.

finite
NA1.5

A finite sequence has a definite beginning and end.

flow chart
NA1.5

A flow chart is a diagram that describes a sequence of operations.

formula, formulae
A2.5, A2.6, A4.5, A4.6, P1.3

A formula is a statement that links variables.

fraction
A3.7, N4.1

A fraction is a way of describing a part of a whole.
For example, $\frac{2}{5}$ of the shape shown is red.

frequency
D2.1

Frequency is the number of times something occurs.

frequency diagram
D2.2

A frequency diagram uses bars to display grouped data. The height of each bar gives the frequency of the group, and there is no space between the bars.

frequency table
D2.2

A frequency table shows how often each event or quantity occurs.

function
A2.5, A3.1, A3.2, A4.5

A function is a rule.
For example, $+\,2$, $-\,3$, $\times\,4$ and $\div\,5$ are all functions.

function machine
A3.1, A3.2, A4.1

A function machine links an input value to an output value by performing a function.

general term
NA1.6

The general term of a sequence is an expression which relates its value to its position in the sequence.

generalise
NA1.6

Generalise means formulate a general statement or rule.

generate
NA1.5

Generate means produce.

gradient, steepness
A3.3, A3.5, A5.5, A5.6, A5.7

Gradient is the measure of the steepness of a line.

graph
A3.2, A3.4, A3.6, A5.5,
A5.6, P1.5

A graph is a diagram that shows a relationship between variables.

greater than (>)
NA1.1

Greater than means more than.
For example, $4 > 3$.

grid
S3.2

A grid is a repeated geometrical pattern used as a background to plot coordinate points. It is usually squared.

hectare
N4.6

A hectare is a unit of area equal to 10 000 m².

height, high
S2.1

Height is the vertical distance from the base to the top of a shape.

highest common factor (HCF)
NA1.3

The highest common factor is the largest factor that is common to two or more numbers.
For example, the HCF of 12 and 8 is 4.

horizontal
A4.1

Horizontal means flat and level with the ground.

hundredth
NA1.1

A hundredth is 1 out of 100.
For example, 0.05 has 5 hundredths.

hypothesis, hypotheses
D2.6, D3.1, D3.4

A hypothesis is a statement that has not been shown to be true or untrue.

imperial unit: foot, yard, mile, pint
S2.2

Imperial units are the units of measurement historically used in the UK and other English-speaking countries.

impossible
D1

An event is impossible if it definitely cannot happen.

improper fraction
N2.1

An improper fraction is a fraction where the numerator is greater than the denominator. For example, $\frac{8}{5}$ is an improper fraction.

increase, decrease
N2.5

Increase means make greater. Decrease means make less.

independent variable
P1.2, P1.3

An independent variable does not depend on another variable for its value.

index, indices
A2.2

The index of a number tells you how many of the number must be multiplied together. When a number is written in index notation, the index or power is the raised number
For example, the index of 4^2 is 2. The plural of index is indices.

index notation
A2.2

A number is written in index notation when it is expressed as a power of another number. For example, 9 in index notation is 3^2.

infinite
NA1.6

An infinite sequence has no definite end.

input, output
A3.1, A4.5

Input is data fed into a machine or process. Output is the data produced by a machine or process.

integer
NA1.1, NA1.2, N2.3

An integer is a positive or negative whole number (including zero). The integers are: ..., $^-3$, $^-2$, $^-1$, 0, 1, 2, 3, ...

intercept
A3.3, A5.6

Two lines intercept at the place where they cross. This is called the point of intersection.

interest
N2.4, N2.5, N2.6

Interest is the amount paid by someone who borrows money. Interest is calculated as a percentage of the sum borrowed.

interior angle
S1.2

An interior angle is inside a shape, between two adjacent sides.

interpret
A3.6, D2.3, A5.7, P1.2, P1.4

You interpret data whenever you make sense of it.

interrogate
D3.1, D3.2

To interrogate is to ask questions.

intersect, intersection
S4.9

Two lines intersect at the point, or points, where they cross.

intersection

interval
D3.4

An interval is the size of a class or group in a frequency table.

inverse
N3.8, A4.1, N4.2

An inverse operation has the opposite effect to the original operation. For example, multiplication is the inverse of division.

isometric
S4.1

Isometric grids are designed to make it easier to draw shapes.

justify
D2.6

To justify is to explain or to prove right.

length: millimetre, centimetre, metre, kilometre; mile, foot, inch
S2.1, S2.2

Length is a measure of distance. It is often used to describe one dimension of a shape.

less than (<)
NA1.1

Less than means smaller than.
For example, 3 is less than 4, or $3 < 4$.

likelihood
D1

Likelihood is the probability of an event happening.

likely
D1

An event is likely if it will happen more often than not.

line

A line joins two points and has zero thickness.

line graph
D3.4

Points representing frequencies are joined by straight lines on a line graph.

line of symmetry
S3.2, S3.4

A line of symmetry is a line about which a 2-D shape can be folded so that one half of the shape fits exactly on the other half.

line symmetry
S3.2, S3.4

A shape has line symmetry if it has a line of symmetry.

linear equation, linear expression, linear function, linear relationship
A2.3, A3.1, A4.2, A4.3, A5.6

An equation, expression, function or relationship is linear if the highest power of any variable it contains is 1.
For example, $y = 3x - 4$ is a linear equation.

linear sequence
NA1.6

The terms of a linear sequence increase by the same amount each time.

locus, loci
S4.7, S4.8

A locus is the position of a set of points, usually a line, that satisfies some given condition. Loci is the plural of locus.

lowest common multiple (LCM)
NA1.3, N2.2

The lowest common multiple is the smallest multiple that is common to two or more numbers.
For example the LCM of 4 and 6 is 12.

lowest terms
N2.1

A fraction is in its lowest terms when the numerator and denominator have no common factors.

map
S3.2

To map is to follow the rule of a mapping to link two sets of numbers.

mapping
A3.1

A mapping is a rule that can be applied to a set of numbers to give another set of numbers.

mass: gram, kilogram; ounce, pound
S2.1, S2.2

Mass is a measure of the amount of matter in an object. An object's mass is closly linked to its weight.

mean
D2.5, D3.3

The mean is an average value found by adding all the data values and dividing by the number of pieces of data.

measure
S2.1

When you measure something you find the size of it.

median
D2.4, D2.5, D3.3

The median is an average which is the middle value when the data is arranged in size order.

midpoint
S1.5, S4.6

The midpoint of a line is halfway between two points.

mirror line
S3.2, S3.3

A mirror line is a line or axis of symmetry.

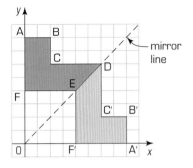

mixed number
N2.1

A mixed number has a whole number part and a fraction part. For example, $3\frac{1}{2}$ is a mixed number.

modal class
D2.4, D3.3

The modal class is the most commonly occurring class when the data is grouped. It is the class with the highest frequency.

mode
D2.4, D2.5, D3.3

The mode is an average which is the data value that occurs most often.

model
D4.1

A model is a formula that attempts to copy a real-life pattern.

multiple
NA1.3

A multiple of an integer is the product of that integer and any other. For example, these are multiples of 6: $6 \times 4 = 24$ and $6 \times 12 = 72$.

multiply, multiplication
NA1.2, N3.4, N3.5, N4.2, N4.4

Multiplication is the operation of combining two numbers or quantities to form a product.

multiply out (expressions)
A4.4

A bracket is multiplied out when each term inside it is multiplied by the term outside it.
For example, $3(x + 1)$ multiplied out is $3x + 3$.

mutually exclusive
D1

Mutually exclusive events cannot both occur in one experiment. For example, if you toss a coin once, you cannot get a Head and a Tail.

nearest
NA1.1

Nearest means the closest value.

negative
NA1.1, NA1.2, N4.1, N4.2

A negative number is a number less than zero.

net
S2.6

A net is a 2-D arrangement that can be folded to form a solid shape.

nth term
NA1.6

The nth term is the general term of a sequence.

numerator
N2.2

The numerator is the top number in a fraction. It shows how many parts you are dealing with.

object, image
S3.2, S3.3, S3.5

The object is the original shape before a transformation. An image is the position of the object after a transformation.

operation
NA1.1, N3.1, N3.6

An operation is a rule for processing numbers or objects. The basic operations are addition, subtraction, multiplication and division.

opposite (sides, angles)
S3.1

Opposite means across from.

The red side is opposite the red angle.

order
NA1.1

To order means to arrange according to size or importance.

order of magnitude
N3.6, N4.5

Order of magnitude is approximate measure of size. It is used when estimating answers.

order of operations
N4.3, A4.6

The conventional order of operations is:
brackets first,
then division and multiplication,
then addition and subtraction.

order of rotational symmetry
S3.4

The order of rotational symmetry is the number of times that a shape will fit on to itself during a full turn.

origin
A3.2

The origin is the point where the x- and y-axes cross, that is $(0, 0)$.

outcome
D1.1, D1.2

An outcome is the result of a trial or experiment.

parallel
S1.1

Two lines that always stay the same distance apart are parallel. Parallel lines never cross or meet.

partition, part
A2.4, N3.3, N3.4, N3.5

To partition means to split a number into smaller amounts, or parts. For example, 57 could be split into $50 + 7$, or $40 + 17$.

percentage (%)
N2.4, N2.5, N2.6

A percentage is a fraction expressed as the number of parts per hundred.

perimeter
A5.1, A5.8

The perimeter of a shape is the distance around it. It is the total length of the edges.

perpendicular
S1.5, S1.6, S2.3, S2.4

Two lines are perpendicular to each other if they meet at a right angle.

perpendicular bisector
S4.7

The perpendicular bisector of a line is at right angles to the line at its midpoint.

pie chart
D2.2, D2.3

A pie chart uses a circle to display data. The angle at the centre of a sector is proportional to the frequency.

place value
NA1.1

The place value is the value of a digit in a decimal number. For example, in 3.65 the digit 6 has a value of 6 tenths.

plan
D2.1

A plan is an overview.

plan view
S4.2

A plan view of a solid is the view from directly overhead.

polygon: pentagon, hexagon, octagon
S1.2, S3.4, P1.1

A polygon is a closed shape with three or more straight edges.

A pentagon has five sides.

A hexagon has six sides.

An octagon has eight sides.

population
D2.3

The entire group from which a sample is taken.

population pyramid
D2.3

A population pyramid is a back-to-back bar chart showing the differences between two populations.

position-to-term rule
NA1.6

The position-to-term rule links the value of a term to its position in the sequence.

positive
NA1.1, NA1.2, N4.1, N4.2

A positive number is greater than zero.

power, index, indices
A2.2, NA1.4, N3.1, N4.3, N4.6

When a number is written in index notation, the power or index is the raised number.
For example, the power of 3^2 is 2.

power key
N3.1

The power key on a calculator is used to calculate a number raised to a power.

predict
D3.4

Predict means forecast in advance.

primary (data)
D2.1, D3.1

Data you collect yourself is primary data.

prime
NA1.3

A prime number is a number that has exactly two different factors.

prime factor decomposition
NA1.3

Expressing a number as the product of its prime factors is prime factor decomposition.
For example, $12 = 2 \times 2 \times 3 = 2^2 \times 3$.

probability
D4.1, D4.2, D4.3

Probability is a measure of how likely an event is.

probability scale
D1

A probability scale is a line numbered 0 to 1 or 0% to 100% on which you place an event based on its probability.

product
N3.1

The product is the result of a multiplication.

proof, prove
S1.1, S1.2

A proof is a chain of reasoning that establishes the truth of a proposition.

proportion
N2.6, N3.7, N3.9, A5.5, P1.4

Proportion compares the size of a part to the size of a whole. You can express a proportion as a fraction, decimal or percentage.

protractor (angle measurer)
S1.2

A protractor is an instrument for measuring angles in degrees.

quadrant
S4.6

A coordinate grid is divided into four quadrants by the *x*- and *y*-axes.

quadrilateral: kite, parallelogram, rectangle, rhombus, square, trapezium
S1.2, S1.4, S2.3, S2.4

A quadrilateral is a polygon with four sides.

rectangle

All angles are right angles. Opposite sides equal.

parallelogram

Two pairs of parallel sides.

kite

Two pairs of adjacent sides equal. No interior angle greater than 180°.

rhombus

All sides the same length. Opposite angles equal.

square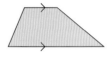

All sides and angles equal.

trapezium

One pair of parallel sides.

questionnaire
D3.2

A questionnaire is a list of questions used to gather information in a survey.

quotient
N3.6

A quotient is the result of a division.
For example, the quotient of 12 ÷ 5 is $2\frac{2}{5}$, or 2.4.

random
D1.5

A selection is random if each object or number is equally likely to be chosen.

range
D2.4, D2.5, D3.3

The range is the difference between the largest and smallest values in a set of data.

ratio
N3.7, N3.8, N3.9, S3.5, A5.5
P1.4, S4.4

Ratio compares the size of one part with the size of another part.

recurring decimal
N2.1, N3.2

A recurring decimal has an unlimited number of digits after the decimal point.

reflect, reflection
S3.2, S3.3

A reflection is a transformation in which corresponding points in the object and the image are the same distance from the mirror line.

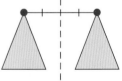

reflection symmetry
S3.4

A shape has reflection symmetry if it has a line of symmetry.

regular
S3.4

A regular polygon has equal sides and equal angles.

relationship
A3.1

A relationship is a link between objects or numbers.

remainder
N3.6

A remainder is the amount left over when one quantity is exactly divided by another. For example, 9 ÷ 4 = 2 remainder 1.

represent
D2.2

You represent data whenever you display it in the form of a diagram.

rotate, rotation
S3.2, S3.3

A rotation is a transformation in which every point in the object turns through the same angle relative to a fixed point.

rotational symmetry
S3.4

A shape has rotational symmetry if when turned it fits onto itself more than once during a full turn.

round
N3.2

You round a number by expressing it to a given degree of accuracy. For example, 639 is 600 to the nearest 100 and 640 to the nearest 10.
To round to one decimal place means to round to the nearest tenth. For example 12.47 is 12.5 to 1 dp.

rule
P1.6

A rule describes the link between objects or numbers. For example, the rule linking 2 and 6 may be +4 or ×3.

ruler
S2.1

A ruler is an instrument for measuring lengths.

sample
D2.1, D3.1, D3.2

A sample is part of a population.

sample space (diagram)
D1.1, D1.2, D1.3

A sample space diagram records the outcomes of two events.

scale, scale factor
N3.8, N3.9, S3.5, S3.6

A scale gives the ratio between the size of the object and its diagram. A scale factor is the multiplier in an enlargement.

scale drawing
S4.4

A scale drawing of something has every part reduced or enlarged by the same amount, the scale factor.

scatter graph
D2.2, D3.4

A scatter graph is a graph on which pairs of observations are plotted.

secondary (data)
D2.1, D3.1

Data already collected is secondary data.

sequence
NA1.5

A sequence is a set of numbers or objects that follow a rule.

service charge
D2.5

The cost for providing the service.

shape
S1.4, S4.1

A shape is made by a line or lines drawn on a surface, or by putting surfaces together.

side (of 2-D shape)
S4.2

A side is a line segment joining vertices.

sign
NA1.2

A sign is a symbol used to denote an operation.

sign change key
NA1.1, NA1.2

The sign change key +/– on a calculator changes a positive value to negative or vice versa.

significant
NA1

The first non-zero digit in a number is the most significant figure. For example, the most significant figure in 207 is the 2, which represents 200.

simplest form
NA2.1

A fraction (or ratio) is in its simplest form when the numerator and denominator (or parts of the ratio) have no common factors. For example, $\frac{3}{5}$ is expressed in its simplest form.

simplify
A2.3, A2.4, A4.2

To simplify an expression you gather all like terms together into a single term.

simulation
D1.6, D4.1, D4.2, D4.3

A simulation is a mathematical model.

sketch
A3.2

A sketch shows the general shape of a graph or diagram.

slope
A5.7

A slope or gradient is the measure of steepness of a line.

solid (3-D) shape: cube, cuboid, prism, pyramid, square-based pyramid, tetrahedron
S2.5, S2.6, S4.1, S4.3

A solid is a shape formed in three-dimensional space.

cube

six square faces

cuboid

six rectangular faces

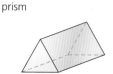
prism

the end faces are constant

pyramid

the faces meet at a common vertex

tetrahedron

all the faces are equilateral triangles

square-based pyramid

the base is a square

solution (of an equation)
A5.4

The solution of an equation is the value of the variable that makes the equation true.

solve (an equation)
A4.3, A4.4, A5.2, A5.3, A5.8

To solve an equation you need to find the value of the variable that will make the equation true.

spin, spinner
D1

A spinner is an instrument for creating random outcomes, usually in probability experiments.

square number, squared
NA1.4

If you multiply a number by itself the result is a square number. For example 25 is a square number because $5^2 = 5 \times 5 = 25$.

square root
NA1.4

A square root is a number that when multiplied by itself is equal to a given number. For example $\sqrt{25} = 5$, because $5 \times 5 = 25$.

statistic, statistics
D1, D2, D3, D4

Statistics is the collection, display and analysis of information.

stem-and-leaf diagram
D2.4, D2.5

A stem-and-leaf diagram is a way of displaying grouped data.
For example, the numbers 29, 16, 18, 8, 4, 16, 27, 19, 13 and 15 could be displayed as:

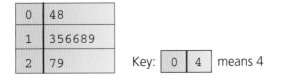

0	48
1	356689
2	79

Key: | 0 | 4 | means 4

straight edge
S1.5, S1.6, S4.8, S4.9

A ruler.

straight-line graph
A3.3

When coordinate points lie in a straight line they form a straight-line graph. It is the graph of a linear equation.

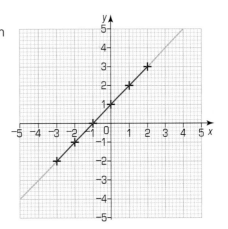

substitute
A2.1, A2.5, A4.6

When you substitute you replace part of an expression with a value.

subtract, subtraction
N3.3

Subtraction is the operation that finds the difference in size between two numbers.

sum
NA1.1

The sum is the total and is the result of an addition.

supplementary angles
S1.2

Supplementary angles add up to 180°.
For example, 60° and 120° are supplementary angles.

surface, surface area
A2.6, S2.6, S4.3

The surface area of a solid is the total area of its faces.

survey
D2.1

A survey is an investigation to find information.

symbol
A2.1

A symbol is a letter, number or other mark that represents a number or an operation.

symmetrical
S3.2, S3.4

A shape is symmetrical if it is unchanged after a rotation or reflection.

T(*n*)
NA1.5

T(*n*) is the notation for the general, *n*th, term of a sequence.
For example, T(3) is the third term.

table
P1.1

A table is an arrangement of information, numbers or letters usually in rows and columns.

tally
D1.4, D2.1

You use a tally mark to represent an object when you collect data. Tally marks are usually made in groups of five to make it easier to count them.

tax
N2.5

Taxes are paid to the government.

temperature: degrees Celsius, degrees Fahrenheit
NA1.1

Temperature is a measure of how hot something is.

tenth
NA1.1

A tenth is 1 out of 10 or $\frac{1}{10}$.
For example 0.5 has 5 tenths.

term
NA1.5

A term is a number or object in a sequence. It is also part of an expression.

terminating decimal
N2.1

A terminating deciamal has a limited number of digits after the decimal point.

tessellate, tessellation
S1.3

A tessellation is a tiling pattern with no gaps. Shapes will tessellate if they can be put together to make such a pattern.

theoretical probability
D1.1, D1.5, D1.6

A theoretical probability is worked out without an experiment.

term-to-term rule
NA1.6

The term-to-term rule links a term in a sequence to the previous term or terms.

theory
D1.6

A theory is a collection of ideas explaining something.

thousandth
NA1.1

A thousandth is 1 out of 1000 or $\frac{1}{1000}$.
For example, 0.002 has 2 thousandths.

three-dimensional (3-D)
S2.5, S4.1

Any solid shape is three-dimensional.

to the power of n
A2.2

This is the index in a number expressed in index notation in general form, for example x^n.

tonne
N4.6

A tonne is 1000 kg.

total
NA1.1

The total is the result of an addition.

transform
A4.2 A4.3, A4.4

A shape is transformed when it is moved from one place to another.

transformation
S3.2, S3.3

A transformation moves a shape from one place to another.

translate, translation
S3.2, S3.3

A translation is a transformation in which every point in an object moves the same distance and direction. It is a sliding movement.

trial
D1.1, D1.4, D4.2

A trial is an attempt to get the right answer.
It is usually followed by an improvement.

triangle: equilateral, isosceles, scalene, right-angled
S1.3, S1.4, S2.3, S2.4, S5.8

A triangle is a polygon with three sides.

equilateral

three equal sides

isosceles

two equal sides

scalene

no equal sides

right-angled

one angle is 90°

triangular number
A2.6

The triangular numbers form the sequence 1, 3, 6, 10, 15, 21, 28 ...
They are the number of dots in a triangular pattern.

triangular prism
S4.3

A triangular prism has a triangular cross-section all
the way through.

two-dimensional (2-D)
S2.3

A flat shape has two dimensions, length and width or base and
height.

two-way table
P1.5, P1.6

A two-way table has two
independent variables along
each side. For example, the
result when you toss two dice
and add the scores.

		Dice 1					
		1	2	3	4	5	6
Dice 2	1	2	3	4	5	6	7
	2	3	4	5	6	7	8
	3	4	5	6	7	8	9
	4	5	6	7	8	9	10
	5	6	7	8	9	10	11
	6	7	8	9	10	11	12

unit fraction
N2.3

A unit fraction has 1 as the numerator.
For example, $\frac{1}{2}$, $\frac{1}{7}$, $\frac{1}{23}$.

unitary method
N2.4, N3.9

In the unitary method you first work out the size of a single unit and
then scale it up or down.

unknown
A2.1, A2.5, A2.6

An unknown is a variable. You can often find its value by solving an
equation.

value
A2.1, A2.5, A2.6

The value is the amount an expression or variable is worth.

value added tax (VAT)
N2.5

A tax imposed by the government.

variable
A2.1, A2.5, A2.6

A variable is a symbol that can take a range of values.

verify
P1.3

To verify something is to show it is true.

vertex, vertices
S2.5, S3.6, S4.1, S4.6

A vertex of a shape is a point at which two or more edges meet.

vertex

vertical
A4.1

Vertical means straight up and down.

vertically opposite angles
S1.1

When two straight lines cross they form two pairs of equal angles called vertically opposite angles.

$a = c \quad b = d$

view
S4.2

The plan view of a shape is the view from above.

volume: cubic millimetre, cubic centimetre, cubic metre
S2.1, S2.5, S4.3

The volume of an object is a measure of how much space it occupies.

whole
NA1.1

The whole is the full amount.

width
S1.4

Width is a dimension of an object describing how wide it is.

x-axis, y-axis
A3.2, S3

On a coordinate grid, the x-axis is usually the horizontal axis and the y-axis is usually the vertical axis.

$(^-2, ^-3)$

x-coordinate, y-coordinate
A3.2, S3

The x-coordinate is the distance along the x-axis.
The y-coordinate is the distance along the y-axis.
For example, ($^-2$, $^-3$) is $^-2$ along the x-axis and $^-3$ along the y-axis.

zero
NA1.1

Zero is nought or nothing.
A zero place holder is used to show the place value of other digits in a number. For example, in 1056 the 0 allows the 1 to stand for 1 thousand. If it wasn't there the number would be 156 and the 1 would stand for 1 hundred.

Thousands | Hundreds | Tens | Units
1 | 0 | 5 | 6

NA1 Check in

1 0.245, 0.246, 0.25, 0.3
2 **a** 30, 60, 20   **b** 30, 60, 15
   c 60, 20   **d** 30, 60, 15, 20
3 **a** Starts at 2, goes up by 3
   b Starts at 3, doubles each time

NA1 Check out

1 **a** ⁻418   **b** ⁻68   **c** ⁻168   **d** ⁻12
2 **a** $2^2 \times 3 \times 5 \times 7$ **b** $2 \times 3^2 \times 5 \times 7$
3 **a** 3, 5, 7, 9   **b** Add 2   **c** 11, 15, 21
4 $2n + 1$

S1 Check in

2 $a = 50°$, $b = 40°$
3 **a** $a = 60°$ **b** $b = 45°$
   c $a = 55°$ **d** $a = 70°$

S1 Check out

1 **a** c and f, d and e
   b a and d, b and c, e and h, f and g
   c a and e, b and f, c and g, d and h
2 **a** ✳ = △   (alternate angles)
      ▢ = ?   (alternate angles)
      △ + ○ + ? = 180°   (angles on a straight line)
      Substituting:
      △ + ○ + ? = ✳ + ○ + ▢ = 180°
   b Split the quadrilateral into two triangles.
      The angle sum of each triangle is 180°,
      so the angle sum of the quadrilateral is
      180° + 180° = 360°
4 A reflex angle is more than 180°, so the
   angle sum would be more than 180°, which
   is not possible.

2

$$0 \quad 0.1 \qquad\qquad 0.5 \qquad\qquad 0.9 \quad 1$$

unlikely   even   highly

    chance   likely

3 a $\frac{1}{6}$ **b** $\frac{1}{2}$ **c** $\frac{1}{2}$

1 $\dfrac{97}{300}$

2

		Dice					
		1	**2**	**3**	**4**	**5**	**6**
Spinner	**1**	1	2	3	4	5	6
	2	2	4	6	8	10	12
	3	3	6	9	12	15	18
	4	4	8	12	16	20	24

3 $\frac{37}{50}$ or 0.74

4 A series of spins should give equal frequencies for all numbers one to five. The spinner should be spun a large number of times to give a good sample size.

1 $\frac{2}{6}$ or $\frac{1}{3}$

2 $\frac{13}{21}$

3 a $\frac{4}{5}$ **b** $\frac{1}{4}$ **c** $\frac{2}{3}$ **d** $1\frac{3}{10}$

4 a 68.4 cm   **b** 96 sheep   **c** 6 minutes

1 a i $\frac{7}{24}$ **ii** $29\frac{1}{6}\%$

   b Steve Stupid (73% compared with 71%)

   c Boys

2 a 123.12 cm   **b** 160.16 cm

   c £10625

3 a $1\frac{22}{35}$ **b** For example: $\frac{1}{3} + \frac{1}{6} + \frac{1}{18}$

1 a $\dfrac{x-6}{2}$ **b** $2(y + 5)$

2 a $2x + 2y$ **b** $4x$

3 a $^-2$ **b** 20 **c** $^-4$

  d 3

1 a $3u - 2t,\ 9u - 6t = 3(3u - 2t)$

2 a $3x + 12$ **b** $xy + xz$ **c** $12t - 8$

  d $17 - x$ **e** $11x - 3 - 5z$

3 a $^-7$ **b** 9 **c** $^-24$

  d 10 **e** 1 **f** 2

4 a

Pattern number	1	2	3	4
Number of squares	1	4	9	16
Number of extra squares	1	3	5	7

  c

1 a m **b** mm

2 a $15\ \text{m}^2$ **b** $22.6\ \text{cm}^2$

3 5 faces, 9 edges, 6 vertices

1 a $6\ \text{cm}^2$ **b** $40\ \text{cm}^2$

  c $13.5\ \text{cm}^2$ **d** $20\ \text{m}^2$

2 a Volume = length × width × height

  b $3840\ \text{cm}^3$

3 Volume = $61.7\ \text{m}^3$, Surface area = $110\ \text{m}^2$

1   6, 15, ⁻9
2   a $y = 0$ b $y = 16$
3   Scalene triangle

1   a

x	⁻3	⁻2	⁻1	0	1	2	3
$y = 3x - 2$	⁻11	⁻8	⁻5	⁻2	1	4	7

2   Difference in intercept:
    $y = 3x - 2$ has intercept of $y = {}^-2$
    $y = 3x + 1$ has intercept of $y = +1$
    Same gradient
3   b x is always one less than y.
    c $y = x + 1$

1   a   ⁻18   b   155   c   ⁻84   d   0.45
2   a   ⁻11.5542   b   ⁻29.65
3   a   3500   b   3480

1   Thierry 60 g, Denis 100 g, Sylvain 120 g
2   £39.48
3   a   56.35   b   30.02   c   2.69   d   2.25
4   a   850 g   b   12.5 g

1 The lines $x = {}^-2$ and $y = 1$.
2 £600, £1000, £400

1 a Coordinates are:
(4, 4), (6, 6), (6, 4), (6, 2)
   b Coordinates are:
(4, 2), (7, 5), (7, 2), (7, ${}^-1$)
2 a i Rotation about (0, 0) through 90° anticlockwise
      ii Rotation about (0, 0) through 180°
      iii Rotation about (0, 0) through 90° clockwise
      iv Translation $\left({}^-{}^{10}_{0}\right)$

1 a $3a$ b $6b$ c $2c^3$ d $3d^2$
2 a $^-1$ b $^-1$ c $^-2$ d 2 e $^-7$

1 $m + m + m + m + n + n + n + n + n + n =$
$4m + 6n = 2(2m + 3n)$
2 a $4x - 4$ b $6x + 10$ c $6 + 2x$
   d $4 - x$
3 $30 = 2(l + 9)$, $l = 6$
4 $2(x + 6) = 18$ $x = 3$
5 $n + (n + 1) + (n + 1 + 1) = 3n + 3 = 3(n + 1)$
which is always divisible by 3:
$$\frac{3(n + 1)}{3} = n + 1$$

D2 Check in

1 Bars with heights 2, 7, 13 and 3

2 Mean = 3.9, median = 4, mode = 5, range = 7

D2 Check out

1 **a** Pie chart with angles: Cat 180°, Dog 90°, Hamster 54°, Gerbil 36°

b Bars with heights 10, 5, 3, 2

c The pie chart shows that half of the pets are cats.
The bar chart shows exactly how many different pets are owned.

d Scatter diagram with points (4, 26), (3, 22), (2.5, 19), (9, 55), (7, 42), (5.5, 30), (4, 32).

e Scatter diagrams are appropriate for data which is not recorded at regular intervals, and for comparing two sets of data.

N4 Check in

1 **a** $^{-}6$ **b** 11 **c** 4.4 **d** 10.5

2 **a** 1.2 **b** 4890 **c** 38.7 **d** 0.0034

3 **a** $1\frac{1}{10}$ **b** $1\frac{9}{14}$ **c** $1\frac{1}{24}$ **d** $\frac{49}{72}$

N4 Check out

1 **a** $40 \times 7 = 280$ **b** $4 \times 30 = 120$

c $160 \div 20 = 8$ **d** $70 \div 20 = 3.5$

e $540 \div 6 = 90$

2 **a** 264.92 **b** 110.78 **c** 6.8

d 4.02 **e** 92.59

3 **a** 2.24

b **i** 22.4 **ii** 2.24 **iii** 0.224

c 37.04

d **i** 3.704 **ii** 37.04 **iii** 3.704 **iv** 37.04

1 a ⁻5 **b** 24 **c** ⁻2 **d** 0
2 a 6 **b** ⁻40 **c** ⁻8 **d** 4
3 a $5x$ **b** $6y^2$
4 $y = x + 1$

1 Vertical: $x + y + x - y + z + x - z = 3x$
$x - y - z + x + x + y + z = 3x$
$x + z + x + y - z + x - y = 3x$

Horizontal: $x + y + x - y - z + x + z = 3x$
$x - y + z + x + x + y - z = 3x$
$x - z + x + y + z + x - y = 3x$

Diagonal: $x + y + x + x - y = 3x$
$x - z + x + x + z = 3x$

2 a $3x - 6 = 5x - 20$ $^-2x = ^-14$ $x = 7$
b $4a + 8 = 6a - 12$ $20 = 2a$ $a = 10$

3 Same gradient (slope), different intercept.

4 $x - 2 = 10 - x$ $2x = 12$ $x = 6$
Sides are length 4, 4 and 11.

5

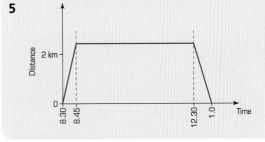

1 a 5, 2, ⁻1
b Starts at 17, goes down by 3
2 a $s + t$ **b** $4s + 3t$ **c** $4s$
3 M-shaped graph

1 Number of hexagons = 3 × Number of squares + 1
2 b Points: (1, 4), (2, 7), (3, 10), (4, 13), (5, 16), ...
3 No
4 Jim

1 7.14 m²
2

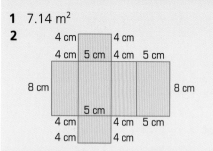

4 cm   4 cm
4 cm   5 cm   4 cm   5 cm
8 cm   8 cm
5 cm
4 cm   4 cm   5 cm
4 cm   4 cm

3 A(2, 6), B(6, 6), C(2, 4), D(6, 4)

1 *abc*
2 **a** Volume = 360 cm³,
Surface area = 324 cm²
   b 8200 cm³
4 (6, 4)
5 No, 4 + 5 < 10

1 61 mm or 6.1 cm
2 Median = 1 car, Mode = 1 car
3 Histogram or frequency polygon with
frequencies: 7, 21, 14, 2

1 **a** Bar chart   **b** Scatter graph
   c Line graph
2 **a** Dates on 10p coins
   b Prices for train journeys of different
lengths
   c Number of newspapers bought over
several years

1 a 0

   b $\frac{8}{20} = \frac{2}{5}$

   c $\frac{7}{20} + \frac{5}{20} = \frac{12}{20} = \frac{3}{5}$

2 $\frac{12}{30} = \frac{2}{5}$

1 a $\frac{8}{24} = \frac{1}{3}$

   b $1 - \frac{4}{24} = \frac{20}{24} = \frac{5}{6}$

2 $\frac{12}{50} = \frac{6}{25}$

3 a $\frac{25}{150} = \frac{1}{6}$

   b $3, \frac{4}{15}$

   c Biased towards 3. More trials would give a clearer result.

Index

Index